THE EXPERIENCE OF NATURE

To the good earth
and all things green and growing

This means acknowledging our kinship with the rest of the biosphere. If we do not feel perfectly at home here, that may after all have something to do with the way in which we have treated the place. Any home can be made uninhabitable. Our culture has too often talked in terms of *conquering* nature. This is about as sensible as for a caddis worm to talk of conquering the pond that supports it, or a drunk to start fighting the bed he is lying on. Our dignity arises *within* nature, not against it.

Mary Midgley (1978, p. 196)

The Experience of Nature

A Psychological Perspective

RACHEL KAPLAN AND STEPHEN KAPLAN
University of Michigan

Ulrich's Bookstore
549 East University Avenue, Ann Arbor, Michigan 48104-2586

Printed in The United States of America

Library of Congress Cataloging-in-Publication Data

Kaplan, Rachel.

The experience of nature : a psychological
perspective / Rachel Kaplan and Stephen Kaplan.

p. cm.

Bibliography: p.

Includes index.

ISBN 0-914004-51-4
1. Nature - Psychological aspects. 2. Environmental psychology.
I. Kaplan, Stephen. II. Title.
BF353.5.N37K37 1989 88-31575
155.9' 15 - dc19 CIP

This book was originally published by Cambridge University
Press in 1989.

CONTENTS

FOREWORD

The natural environment is increasingly a source of interest, fascination, and affection. In a wide range of settings, both at home and abroad, I have found the breadth and intensity of such feelings to be remarkable. Yet until now there has been little material available for individuals who are attempting to understand their feelings toward nature. Though much has been written on this subject, there remains a need for a comprehensive work that is scientifically based, readable, and helpful.

It is clear that, were there such a volume, it would be important to many people. Nature centers, arboreta, and a host of other institutions serve people's need for nature based on individual experience and intuition, but they lack a solidly grounded theoretical framework to guide their efforts. Many individuals are eager for a deeper understanding of the role nature plays in their lives. Gardeners, bird watchers, hikers, nature lovers of many varieties would welcome a way to relate intellectually to these powerful emotional forces.

The Experience of Nature is, at last, such a volume. It establishes a basic understanding of nature experiences from window box to wilderness.

This foreword is addressed particularly to readers who are not trained as scientists. Although the volume is written to serve both scientists and nonscientists, the latter may benefit from a bit of guidance that the former may not require.

When I first started the manuscript I wondered whether it would be too academic to reach the wide audience that could benefit from it. As I continued reading, however, the "plot thickened," and I could see the wisdom of the approach. The scientific-academic basis for the field, clearly established in part I, lays the groundwork for the rest of the book.

Although part I will be challenging for some readers, a detailed understanding of this more theoretical material is not essential to enjoying the book. And, despite the somewhat technical nature of the content, the writing is clear and free of jargon; even difficult concepts are presented in a relaxed, free-flowing style.

Part II presents a more intimate view of nature experiences and is more easily accessible to the layperson. Here the technical concepts of perception and preference give way to a fascinating presentation of human benefits and satisfactions. These insights into human gains, central to the thesis of the book, provide much food for thought. The warm, meaningful responses presented here lead to a deeper understanding of one's self, allowing one to explore one's own nature experiences.

From part II to the end, I felt comfortable and indeed rewarded on every page, rejoicing in seeing the importance of relationships with green nature so clearly established. The book provides an important perspective for looking at one's self journeying through this world, allowing one to examine more closely what had only been suspected, tying personal nature experiences into the larger web of human experience.

I think the volume creates a baseline for anyone interested in the green experience. It provides the scientific basis for inquiry as well as an invitation to think more deeply about this ancient and perhaps too easily accepted role of vegetation as being "good" for people. Everyone who is interested in this relationship will find here an approach that is insightful and widely applicable. They will also find a perspective that they can use as they seek to understand their own nature experiences or to enhance the experiences of others. I am grateful to Rachel and Stephen Kaplan for the exciting and useful synthesis they have achieved in this important volume.

Charles A. Lewis
The Morton Arboretum
Lisle, Illinois

PREFACE

Our work over the last two decades has brought us into contact with an interesting assortment of individuals – landscape architects, horticulturists, foresters, nature interpreters, planners, resource managers, and many more. A common denominator across their diverse fields has been an interest in the effect the natural environment has on people. These individuals have had many questions for us. Three major issues have emerged as being of particular interest. These are:

1. Is it real? Is the effect of nature on people as powerful as it intuitively seems to be?
2. How does it work? What lies behind the power of environments that not only attract and are appreciated by people but are apparently able to restore hassled individuals to healthy and effective functioning?
3. Are some natural patterns better than others? Is there a way to design, to manage, to interpret natural environments so as to enhance these beneficial influences?

We have enjoyed working on these topics, in collaboration with students and colleagues in various disciplines. Gradually, over the years, the outlines of responses to these questions have begun to emerge – not definitive answers by any means, but ways of thinking about these issues that colleagues, students, and many practitioners say they find stimulating and useful. Gradually these ways of thinking have grown and converged. They have come to constitute both a critical mass of material and a framework that has a reasonable degree of coherence.

The time has come to think of making this material available in a more widely sharable form. The issues are not by any means wrapped up and settled. Nonetheless, there are at least two compelling reasons for taking this step. The first is that these many nature-related and nature-influencing activities - designing, planning, managing, interpreting - are going on right now, often based on little theory and even less data. Compared to the well-intended but often severely limited intuitive bases for decision making currently being employed, the results of these years of research and theory could make a positive contribution. Second, we feel the outlines of what is known and what needs to be known have become clear enough that it is time to invite others to join the fun, to participate in the process themselves.

We intend this volume to be of interest to people who may not realize that they have a common bond. Such people include gardeners, backpackers, and bird watchers. Many professionals whose work brings them in direct contact with the natural aspects of the environment have expressed their eagerness to know about the psychological counterpart to the biological knowledge they have acquired. Horticulturists and landscape architects constitute two such professional groups whose training is not as rich in the psychological dimensions as many of them feel it ideally should be.

Yet another intended audience consists of our professional peers in areas such as environmental psychology, environment/behavior research, behavioral geography, environmental education, recreation behavior, and other assorted disciplines. These groups have no common professional association or single home base. We have been fortunate to have many excellent students from these areas in our classes where they have come to learn from each other and to find out that there are strong underlying themes unifying their apparently disparate directions.

Writing for such diverse audiences is a challenge. On the one hand, the topic of the book is widely appealing, and our readers will have great intuitive familiarity with the contents. On the other hand, our intention is to present results of a large number of empirical studies and considerable theoretical material. We have tried to accommodate to these issues and to the differing expectations about supporting information that diverse professional groups hold. Both the chapter notes and the summaries in the Appendixes are intended to provide some of the information that is essential to some more professionally oriented readers but may be distracting to others. There are also references that provide fuller documentation. At the same time, though our intention was to write a readable and interesting volume, we have not attempted a popularized treatment. Our effort is to integrate a substantial quantity of both empirical and theoretical material into a framework and perspective on the role nature plays.

From the present perspective it is difficult to believe that at the time we began this research program there were virtually no studies on the subject of this book. A great deal has happened since then, and the literature on the importance of nature is growing. In retrospect, we find ourselves surprised by the quantity of research we and our students have done in these two decades. The present volume focuses on this more or less coherent research program. Though we refer to the work of others as it seems pertinent, we have placed a higher priority on coherence and integration than on exhaustiveness.

The task of bringing together even this comparatively unified body of research has been a most stimulating challenge. We have learned a great deal in the process. Much as we have been aware of the remarkable power of the natural environment in the lives of people, the convergence of the various lines of research in supporting this conclusion was even more striking than we had realized. The surprising strength of the evidence leads us to hope that those who rise to defend the natural environment against the continuing threats of encroachment will find this volume a useful ally. As an expression of support for this urgent struggle, the royalties from this book are being donated to the Nature Conservancy.

ACKNOWLEDGMENTS

We had little doubt that it would be timely to write the "Nature Book." There was substantial doubt, however, about finding the time to carry out such a major project. Without the support and encouragement of the Bloedel Foundation we would not have begun. The financial support to release some time during the academic year was essential. Even more important, however, was the psychological dimension: an institution that felt our efforts needed to be shared and made the opportunity available.

In this case the institution is also a person. We first met Prentice Bloedel in 1977 through the efforts of Charles

Lewis. Mr. Bloedel's deep-rooted understanding of the importance of nature expressed itself in many ways. The Bloedel Reserve on Bainbridge Island, Washington, is one such expression. We had the good fortune to walk across this land with him as our guide. His brief, modest comments were invariably perceptive. His interest in our work and support of this book are yet a further expression. Our thanks to him are many.

We are also grateful to the U.S. Forest Service for continuous support of our research since 1970. The Forest Service gets its share of criticism for some of the most visible aspects of its work. The support we have enjoyed, though far less visible and on a much smaller scale, has nonetheless amounted to a substantial contribution. Walt Hopkins and Duane Lloyd in the Washington office in the 1970s, Jim Morgan and various successors in the St. Paul office, and particularly John Dwyer, project leader of the Urban Forestry Project for the North Central Forest Experiment Station in Chicago, have given us a different view of the Forest Service. Their support has extended far beyond the financial; they have actually been interested in the substance of our work and have been excellent disseminators of some of its results.

This book represents the efforts of many. The numerous studies featured in the appendixes include the works of former students and colleagues who have shared our interest in the natural environment. Many of these individuals have contributed to the development of our ideas. Many others have also left a strong mark through their critical comments and searching questions.

The book itself benefited from the efforts of several individuals. Three in particular made substantial contributions: Thomas Herzog commented on several chapters; Lisa Bardwell took major responsibility for preparing Appendix B; Janet Talbot, a colleague of long standing, prepared Appendix D and collaborated on Chapter 5.

Our gratitude and indebtedness are far greater than this listing would suggest. We have had the good fortune to work with outstanding, dedicated students, to be constantly barraged by eager, thoughtful individuals who have run across our work, and to have been part of a community of seekers who share our conviction that the natural environment plays an important role in human well-being.

INTRODUCTION: NATURE AND HUMAN NATURE

NATURE is a valued and appreciated part of life. Examples abound. People plant flowers and shrubs and nurture house plants; cities invest heavily in trees; citizens band together to preserve natural settings they have never seen; landscapes for centuries have been the subject of painting and poetry. Nature seems to be important to people. Though substantial sums of money are spent on nature and natural settings, it is hard to justify the role nature plays in rational terms. In fact, people with relatively little money are no less likely than the more affluent to have a splash of colorful flowers in front of their homes. Bond proposals for parks have often passed even when other issues fail. The grief neighbors feel when "their" tree is removed can hardly be explained on economic grounds.

It is no doubt possible to provide alternative explanations for any one of these examples. As a group, however, they provide at least circumstantial evidence that nature is important in itself rather than for some extrinsic reason. Further evidence in support of this hypothesis is provided by several recent studies using special populations. Verderber (1986) has shown that the quality of the view out the window is a significant factor in the recovery of patients in physical medicine and rehabilitation wards of six hospitals. Ulrich (1984) demonstrated that the content of the view is important in hospital patients' recovery from surgery, with nature content contributing to faster recovery. Moore's (1981) study showed a dramatic relationship between inmates' use of health care facilities at a large federal prison and the view from their cell. Those whose view was of other inmates sought health care most often. Of the inmates whose views were of areas outside the prison building, the ones who looked out onto the surrounding farmland sought health care least of all. West's (1986) study further supported these findings in another prison setting.

These carefully designed studies established clear support that nature in itself plays an important role. After all, it would be most unlikely that the patients and inmates were "using" nature for some other goal, such as to achieve status, to gain personal identity, or to establish territory. The purpose of each of these studies, however, was not to explain why nature plays such an important role, but to make it clear that there is a meaningful phenomenon to examine and that there are important implications in terms of basic human needs.

Many of the studies reported in this volume provide further evidence for the importance of nature to people, but that is not our primary purpose. Rather, our intention is to go beyond this, to focus on what nature does, for whom, under what circumstances.

ABOUT NATURE

This book is about the natural environment, about people, and about the relationship between them. It is about things many have known but few have tried to study empirically. It is about things for which there is only a limited vocabulary.

Nature connotes many settings. As is clear already, our use of the word *nature* is intended to be broad and inclusive (although, for the most part, the discussion excludes fauna). The discussion of nature here is not limited to those faraway, vast, and pristine places where there has been little human intervention, or to places designated as "natural areas" by some governmental authority. Nature includes parks and open spaces, meadows and abandoned fields, street trees and backyard gardens. We are referring to places near and far, common and unusual, managed and unkempt, big, small, and in-between, where plants grow by human design or even despite it. We are referring to areas that would often be described as green, but they are also natural when the green is replaced by white or brown or red and yellow.

Nature includes plants and various forms of vegetation. It also includes settings or landscapes or places with plants. Thus the plants and their arrangement in a space, and the juxtaposition of the plants to other elements in the environment, all play a role in our discussion.

The expression *natural environment* is not intended to include only purely natural elements, any more than the *built environment* refers exclusively to constructed elements. Similarly, the contrast between *natural* on the one hand and *urban* and *rural* on the other we find to be unhelpful. Much of our discussion is about the nature that can be found in the urban and in the rural context.

It is clear that whereas the concept of nature is very much part of the human experience the language for discussing it is neither rich nor precise. Although *nature* and *natural environment* as used here are difficult to define adequately, they refer to things and places we have all experienced.

ABOUT PEOPLE

The studies discussed in parts I and II explore different natural settings and diverse facets of the way people relate to such places. Though these studies have been guided by a theoretical perspective, it is even more clearly the case that the theory has been guided by the studies. Each of the chapters deals with both the studies and the explanatory framework. It may be helpful to provide an introduction to this framework. It is a view that considers humans as deeply concerned with information and examines the environment as a vital source of information.

Human functioning depends on information.[1] Much of this information is provided by the immediate environment. There are signs, both verbal (such as a street name) and nonverbal (such as a doorbell), that provide guidance to behavior. There are combinations and arrangements of elements that constantly require deciphering: a group of people standing near eath other, all facing the same way; a long hallway with many doors; a series of cash registers; a group of trees where two paths cross. Some of the information is urgent and requires action; some is made urgent by size, motion, or color and may be difficult to ignore despite being irrelevant to one's current goals.

A great deal of the information that is essential to functioning is already stored in our heads based on previous experience. Such stored information not only makes it possible to assess a current situation but is essential in anticipating what might happen next. Humans can close their eyes and imagine, they can consider alternate plans, they can give advice or instructions – all based on information that is not immediately present in the environment.

It is clear from this discussion that information is not only what scholars consume and intellectuals exchange. In fact, it is reasonable to say that information (or knowledge) is the common human coin. The storage and processing of information are the cornerstone of human functioning. In addition to relying on the information that has already been acquired, humans are active in their pursuit of additional information. In fact, they often seek information even when having it makes little discernible difference. Indeed, the interest in useless information is so powerful that, according to Postman (1985) in his provocative *Amusing ourselves to death,* it is used against us by the media. He considers television, newspapers, and Trivial Pursuit to be striking examples of this unfortunate turn of events.

People are particularly aware of information that is visual, that concerns what they see. That does not mean that people interpret the information in visual terms exclusively; rather, visual stimuli are effective in conjuring associated information. The sight of water provides information about potential opportunities, which may or may not be visual in themselves. Magazine advertisements rely upon the reader's imagination (and prior experience) to recognize other aspects of the scene than the visual material that is presented. Visual material is thus particularly effective in evoking other kinds of information that had previously been associated with the presented information. (The expression *I see* is an interesting manifestation of the dominance of the visual mode without the necessity that it refer exclusively to visual information.)

Similarly, humans are strongly oriented to spatial information. A nearby moving object (for example, a person passing one's office door) is difficult to ignore. A great deal of information that is not necessarily spatial is, in fact, coded in spatial terms. There are many examples of this in the language (e.g., reference to higher orders, bottom line, feeling high or low or under it). A great deal of problem solving and thinking involves arraying pieces of the problem in a real or imaginary space. Thus, both conceptually and with respect to information in the current environment, people are sensitive to relative location.

One more item is essential to discuss in our brief overview of the theoretical perspective that guides this book. We have said that information is essential to human functioning, that people store and use huge amounts of information, and that they actively seek more information. None of this speaks to the strong feelings people have about information. The relationship we humans have to information is, in many situations, far from neutral. We assess current and future situations in terms of whether they are/will be good or bad, pleasant or painful. The anticipation of an unpleasant situation, even if one is currently in a pain-free situation, can make one's feelings negative. Similarly, people whose conditions are extremely painful can feel much better if they anticipate a hopeful future.

From this perspective, then, people not only are adroit in their use of information but crave it and continuously evaluate it. The evaluation of information is essential; effective functioning relies on sorting the good from the bad, the useful from the useless. Humans judge situations with such facility that they are often not aware of the fact that such an evaluation is occurring. The rapid intuitive evaluation by people of other people is a widely recognized phenomenon; the evaluation of environments occurs with comparable unobtrusiveness and comparable facility.[2]

ORIENTATION TO THE BOOK

This book is a reflection of an odyssey. We have been interested in the relationship between people and the natural environment for nearly 20 years. During that time the varied interests of students we worked with and the diversity of research opportunities have left their mark. As a result we have become involved in a rather wide spectrum of topics in this area, including scenic assessment, gardening, wilderness recreation, and what we have come to call "nearby nature." At many points along the way these various studies did not seem particularly related, but gradually, from these separate parts, a coherent sense of the whole has begun to emerge. As a reflection of this continuing odyssey, this volume is based largely on our own work and that of our students (although the work of others is, of course, referred to as well). Even with respect to our work, this volume is an attempt not to be exhaustive but to emphasize the larger conceptual themes and to bring in specific studies as a means of making these themes more concrete and understandable.

There are three major sections to the book. The last of these, part III, provides a synthesis of what came before. In it we develop the concept of a restorative environment – an environment in which the recovery of mental energies and effectiveness is enhanced. The examination of factors that make it restorative draws on the issues of preferences, satisfactions, and fascinations, the foci of the first two sections.

The first major section of the book, part I, is devoted to research on the perception of and preference for natural settings. Given a species that is strongly oriented to visual and spatial information, and is quick to have feelings about things, what role does nature play? Are there natural elements and natural settings to which people react strongly, and do such reactions show consistent patterns? A particularly important aspect of this section is an understanding of the role that preference plays. Rather than being frivolous, preference as we have come to see it through numerous studies is an expression of a deep and underlying aspect of human functioning. Although people are generally not aware of it, their preferences are influenced by such factors as whether they could learn more in a given environment and whether they would be able to move around in the environment without fear of getting lost.

Consistency in the pattern of preference for natural settings does not, however, suggest universality in what people prefer. Part I also explores variations on the theme – different ways in which such preferences express themselves. Although in general there are broad areas of agreement as to what people like in the natural environment, there are at the same time significant and consistent differences. These are interesting both for what they add to the understanding of preference in general and for their implications for the design and management of natural settings.

Part II deals with research on the satisfactions and benefits people derive from contact with natural environments. Research we have carried out in a wilderness context over a period of 10 years provided a laboratory for studying such satisfactions. Fascination appears as an important component. Compatibility, the match between the demands of the environment and the goals and intentions of the individual, is a particularly strong factor in the wilderness context.

Similar kinds of satisfactions and benefits to those derived in the wilderness are also evident in the immediate natural setting. The nearby natural environment is vital both to how satisfied people are with where they live and to their very quality of life. Research on the satisfactions people obtain from gardening points to the central role of fascination here, too. In the presence of such fascination people are able to rest their minds in a way that allows them to recover their effectiveness.[3]

NOTES

1. Although the concept of information is central to our approach, our version of the information-processing position is relatively unconventional. It views the processing of information in biological systems as different in many respects from the way computers process information. It also takes several steps in the direction of Gibson (1979), an outspoken critic of the information-processing approach. Nonetheless, compared to most other approaches within environmental psychology, our emphasis on the role of information and the importance of tracing its flow through the organism places us at least on the fringe of the information-processing approach. For an exposition of this perspective, see *Cognition and environment: Functioning in an uncertain world* (S. Kaplan & Kaplan, 1982).
2. Implicit in this discussion is the assumption that much of human functioning is impacted by its evolutionary origins. We have chosen not to emphasize this relationship in this volume since the theory and data presented here speak for themselves and are not dependent on this evolutionary background. Readers interested in exploring this potentially enriching theoretical linkage may wish to look at the pertinent portions of *Humanscape* and *cognition and environment* (S. Kaplan & Kaplan, 1978, 1982).
3. In addition to these three major sections the book has four appendixes. These include summaries of a substantial number of studies that are the grist for the material in parts I and II. Rather than interrupt the flow of discussion, we have marked these studies with an asterisk when they are mentioned in the text, which indicates that they are described in the appendixes. Although some readers will find little reason to turn to any of this material, others will find it useful for more detailed study of the empirical background behind the ideas presented here.

PART I

THE PREFERENCE FOR NATURE

THERE is the story of the fellow who is sitting at his rather bare desk when the phone rings. He picks up the receiver and responds to the caller's query, "Yes, we do landscapes." Some seconds later: "Well, whatever you like. How about one with, say, some mountains . . . and a river running through there, and a canoe paddling down the river?" And then, "You don't want a canoe? Oh, that's no problem. Sure. We can do it without. Yeah. OK. We can have that ready for you in a week." As soon as he puts the receiver down, the fellow is heard calling to the back room, "Hey Joe, one landscape and hold the canoe."

Though the canoe may be optional, the presence of water is highly likely in a made-to-order preferred landscape. It can be an ocean, a big lake, a small lake, river, stream, or pond; it might be placid or fast-moving, tranquil or falling, with trees reflected or with rapids. Water is a highly prized element in the landscape.

Water is not just an attractive element in pictures. People are willing to pay higher rents for a water view. In fact, along many bodies of water the density of residential development provides an all too vivid indication of people's willingness to be near the water even if they must share the view, the facilities, the roads, the traffic, and everything else. The fondness for the water seems to hold whether it is a place for active water sports or not, whether one plans to be "using" the water or is unlikely to ever directly interact with it. It turns out that much of the enjoyment that waterscapes provide is low in action.

Water provides an excellent example of an aspect of the natural environment that is highly preferred. Though water seems to be an attractive element, it is also the relationship of the water to its surroundings that is important in the preference. It is, thus, meaningful to examine not only the specific "things" or elements or contents of the natural environment that are preferred but also the relationship among these elements. A landscape is more than the enumeration of the things in the scene. A landscape also entails an organization of these components. Both the contents and the organizational patterns play an important role in people's preference for natural settings.

Before we launch into an examination of preference, it is important to be clear about why this is a worthwhile topic at all. *Preference* has a frivolous connotation. It suggests the decorative rather than the essential, the favored as opposed to the necessary. It also seems idiosyncratic – tastes are known to vary, after all. Given that the world is less than perfect, that many people do not even have their basic needs met, preference may appear to be a luxury that only few can afford to consider.

The view we take of environmental preference is in strong contrast to such a position. Preference is intimately tied to basic concerns. We see preference as an expression of underlying human needs. Preference can be expected to be greater for settings in which an organism is likely to thrive and diminished for those in which it may be harmed or rendered ineffective. Thus humans, like other animals, are far more likely to prefer a setting in which they can function effectively.

Aesthetic reactions thus reflect neither a casual nor a trivial aspect of the human makeup. Rather, they appear to constitute a guide to human behavior that is both ancient and far-reaching. Underlying such reactions is an assessment of the environment in terms of its compatibility with human needs and purposes. Thus aesthetic reaction is an indication of an environment where effective human functioning is more likely to occur.

Such a position does not require that people are necessarily aware of their needs or that preferences be universal. The way preference feels to the perceiver stands in sharp contrast to the process that underlies it. Preference feels direct, immediate, and holistic. One experiences no hint that one is going through a complex, analytic process en route to one's judgment. At the same time, research results point to a variety of variables that must be assessed and integrated in making a preference judgment. Such an underlying process must be both very fast and quite unconscious.

Part I consists of three chapters. The way the environment is perceived or experienced is the focus of the first of these. Perception and preference are closely related. Perception is a key element in preference, and the measurement of preference permits an examination of the perceptual process. Perception is also strongly influenced by previous experience. The same environment may be perceived in distinctly different ways by a planner, a developer, and a neighbor. The emphasis in chapter 1 is on the way the natural environment is experienced by people with no particular training in environmental domains. Even though each of us has had extensive experience in such settings, our awareness of the underlying categories of the experience is at best minimal. Through studies that focus on preference, however, it has been possible to extract what these categories are. These underlying categories of the way one experiences the environment provide considerable insight into what it means to "function effectively."

In chapter 2 the focus is on factors that seem to account for preference of natural settings. These factors include both contents – particular kinds of settings and natural elements (such as water) – and patterns of organization of the natural setting. As it turns out, these factors do not constitute a long list. Rather, they can be understood easily in terms of the human inclination to seek and process information. Thus the relationship between the preference factors and the informational patterns plays a central role in the chapter.

Chapter 3 deals with some variations in the ways these major informational themes and preferences express themselves. Cultural and ethnic patterns can be reflected in distinctly different preferences for kinds of natural settings. There are also other sources of variation in preference based on people's experiences. For example, individuals whose experience includes professional training in design-related fields may show distinctly different preference patterns. Variations are important to identify and understand if one takes the notion of preference seriously. Understanding both the common themes and their variations is essential if environmental decision making is to be responsive and humane.

CHAPTER 1

PERCEPTION AND CATEGORIZATION

APPRAISALS have traditionally been the province of experts. There are experts who evaluate jewels, antiques, homes, paintings, and many other objects. There are also experts who assess the aesthetics of the scenery. What all these individuals have in common are training and experience that enable them to hold a standard against which they compare a particular example that is to be assessed for its value. One gold watch might be worth $300 whereas another is appraised at 10 times as much. A particular section of highway may be designated as scenic whereas other sections fail to achieve such designation. Such evaluations are considered to have objectivity both because they are judged relative to standards and because the appraiser is assumed to make these judgments in terms of established standards rather than personal reactions.

The process of making these appraisals requires that one categorize or classify. To decide that a particular painting is worth a certain amount depends on classifying it in terms of various designations: artist, age, condition, frame, and so forth. To make decisions about the permissibility of changing a setting - for example, in siting a building in a national park - the current setting must be classified in terms of such considerations as its present use, the sensitivity of the area, its uniqueness. Thus, in achieving expert judgments as to the value of an object or setting, it is necessary to make a series of decisions about the categories to which it belongs.

The choice of appropriate categories for making an appraisal depends on the purpose of the assessment, but it is also affected by the expert's area of training. The appraised value of the same object may differ depending on the specialized knowledge of the appraiser. The nature of the appraisal may also differ if the purpose is to set a purchase price or to provide a plan for modifying it. Thus professional advice about changing the grounds around a corporate headquarters may call on factors other than the monetary value of the site but may still require a series of categorizations about the value of what is there now, the kind of setting, who will use it, and so on.

It has been traditional to make a distinction between expert-based appraisals and the preferences that people have for the same objects or places. People's preferences are not guided by special training, nor do they have knowledge about the appropriate standards. For such reasons, preferences – as opposed to appraisals – are thought to be discrepant and subjective. Nonetheless, there are strong similarities between the process used by the expert in making an appraisal and by the untrained individual who is making a preference judgment. For both expert and nonexpert, a great deal of experience is brought to bear on the decision, and categorization is necessarily an aspect of the decision. Neither expert nor nonexpert is likely to be aware of the process or categories that lead to the decision.

This chapter is about the categories that are implicit in the experience of the natural environment for individuals who have no particular training in environmental fields. Although geologists, for example, may be cognizant of the essential categories underlying their work, the underlying categories for the untrained person are unlikely to include bedrock and moraines. But what are the categories? Given the lack of training in appraising the environment, one could expect the categories to be idiosyncratic and as different as people are different. This has not proven to be the case. Rather, we have found strong commonalities in these categorizations. Because people are rarely aware of how they experience an environment, identifying these common themes in the perceptual process requires an indirect approach. As it turns out, preferences provide a means for discovering the categories of perception. Let us begin, then, by describing the research approach that made it possible to study these categories.

THE RESEARCH FRAMEWORK

Back in the time of Earth Day there was growing concern about environmental degradation and increasing consciousness about the ugliness of the physical surroundings. It became apparent to us (trained as experimental psychologists) that in very important respects the environment was ignored in a great deal of psychological research. Research on aesthetics, potentially a pertinent topic, was largely based on how much people liked artificially constructed stimuli - patterns designed to differ in their amount of complexity. Such material has great advantages to the scientist. The issue of expert appraisal could be totally ignored as few people would have extensive experience with such material. Furthermore, by constructing the stimuli the researcher has full control over their properties. And control is, after all, of utmost importance to the scientist.

Knowledge gained about preference for such patterns may have little relationship to what people like and dislike in the world about them. To find out about preference for existing settings, however, one must give up some of the control. Since the purpose of scientific control is to increase confidence that one is, in fact, isolating the effect of particular factors, reduction of such control must also reduce one's confidence. This is an unavoidable trade-off; one cannot have it both ways. The studies we will be discussing flow from the judgment that control over irrelevant material is of less utility than knowledge gained about relevant material, though under less than optimal conditions. The way to gain confidence despite this limitation is by the accumulation of knowledge across numerous studies. If the pattern of results turns out to be consistent despite great diversity in both the settings that are studied and the backgrounds of the study participants, one's confidence is considerably bolstered.

The research program that we started at that time (around 1970) thus focused on scenes of the environment rather than on artificial stimuli. In other respects, however, the research followed previous work in its basic approach. Individuals were asked to look at each stimulus - now a scene - and to indicate how much they liked it. In our early studies the scenes were presented for a fixed time, but later studies - using printed photographs rather than slides - permitted individuals to pace themselves. As was true of earlier psychological research, we also had the scenes rated in terms of other qualities, which could then be related to preference.

Such a procedure permits use of meaningful material

while still maintaining some elements of control. The researcher selects what is to be viewed. The scenes are presented as slides or photographs, thus eliminating some of the experimental "noise" of the actual physical setting. A fairly large number of scenes can be presented using this method. This makes it possible to examine more carefully the factors that affect the preferences.

Once again, as soon as such controls are introduced the trade-off dilemma arises. Would it not be better, or more accurate, to take people to particular settings and have them rate these for preference? After all, photography can be deceiving. One can take pictures so as to make a setting more pleasant than it actually is. The "noise" in the real setting brings in other sense modalities – sound, smell, touch. Are judgments made in response to slides of settings in any sense comparable to those that would be made in the setting itself?

The quick and straightforward answer to these questions is that people's responses to the two-dimensional representation are surprisingly similar to what they are in the setting itself. A number of studies have focused on these issues because these are basic questions that needed to be resolved.[1] Though the similarity in responses to pictures and to real settings may seem surprising to those who focus on research methods, it is much less surprising if considered in terms of daily human experience. Much of the information that we consider all the time reaches us by means of two-dimensional representations of three-dimensional settings. When watching television or seeing pictures in a book or a painting on the wall people are not likely to say that the representation is deceiving.

It would, in fact, be much more difficult to justify a procedure that relied upon preference judgments made in the physical environment itself. The process of bringing the study participants to each of the various settings would in itself introduce a great variety of problems. It would also be difficult to ask the participants to consider as many settings under these circumstances as can easily be used with photographs.

Before we leave these methodological concerns, there is one more that needs to be mentioned. Although our earliest studies used color slides to represent the settings, many of the later studies relied upon small black-and-white printed pictures. This raises the question of people's ability to judge material that is not in color and where the photographic quality is short of outstanding. Again, the quick answer is that this poses no problems. Participants might express regret that the scene is not in color, but their ability to react to it is not hampered. Participants also often "read" color into the scene. Natural settings, for instance, may be seen as green despite the absence of color. Though there certainly are some advantages to color, it has disadvantages as well. Color that is poorly matched to one's expectations (lawns that are too blue, for example) creates greater difficulties than black and white. Furthermore, the use of black-and-white photographs makes it possible to distribute the material in printed form at minimal cost.

The material presented in this chapter is based on research that used photographs of real environments and asked people to judge these for preference. In all cases individuals were asked to rate each scene in the study using a 5-point rating scale (where 1 means they "don't like it" and 5 means they "like it a great deal"). Each of the studies involved various other aspects and focused on a different kind of setting and on a variety of other issues.

A substantial number of studies that have been carried out using this procedure are not discussed in detail within the chapter. Rather, many of them are presented in summary form in Appendix B (and are noted by * when mentioned in the chapter) so that the treatment here can focus on the insights gained from the studies as a whole. Neither the discussion nor the appendix, however, is exhaustive. By now there are more studies that accrue to this line of research than we even know about, and there are others that are sufficiently redundant with what is included that they did not seem needed.

CATEGORIZATION

Professional training involves learning to see things in a particular way. Combinations of elements that had previously not been associated with each other for the novice are given a name that others, trained in the same area, also recognize as salient. The elements might be symptoms that in combination can be recognized as a disease. They may be treatment of doors, windows, and roof lines that combine to be designated as particular architectural styles. Such elements might also be different patterns of cultivation or of land use or of many other things. This learned way of seeing particular combinations of elements is essential to the experts who share it. To call something a mixed stand or a storm drain or a subdivision achieves a shared understanding that carries a great deal of information that can remain unstated because it is implicit in the name. During the brief period when one is acquiring such new information and learning to identify the combination of elements in terms of its new appellation, one may be sensitive to the fact that a new way of seeing is indeed developing. Before long, however, the new terminology is a part of one's vocabulary, and what had been a new way of seeing seems no different than the way one had seen things all along.

When trying to explain some of these newly learned concepts to someone else, one often realizes that the concept has become complex and that the explanation is far from straightforward. Much more than a vocabulary has been acquired. A process of perceptual learning has taken place that includes distinctions that may be difficult to verbalize. Having seen various instances of the new concept leads to a richness of meaning that goes beyond any single example. This very richness had presented difficulties earlier in learning the concept, and now, in trying to communicate the concept to someone who is new to it, the multiplicity of its facets becomes evident.

None of this is very different from the rest of our knowledge and experience.[2] The things we know about, even in areas where we do not consider ourselves to be expert, follow the same patterns. We see things in particular ways and have names for certain combinations of elements; these names apply to a richness of patterns that makes definition difficult. All of us use words like *park* and *garden* and *nature*. We have a shared understanding about them. What constitutes a park? Do parks have swings? benches? walks? No simple definition suffices . We can accept the notion that parks differ and yet they are parks (although it may be more difficult to accept "research parks" and "industrial parks" as included in the same concept). A backyard garden may be like a park in many respects, but we would usually not include those in what we designate a park.

These issues are of utmost importance in considering perception. Though there are categorical distinctions underlying perception, we generally do not realize what these categories might be. Preference judgments, in turn, are based on perceptions. The research on preference thus tries to determine not only what people do and do not like but also what some of the categories are that constitute the basic patterns of daily experience.

Category-Identifying Methodology

The categorizations that experts use in particular settings are often not meaningful (and may even be disturbing) to those who do not share that expertise. The nonexpert "doesn't see it that way." For example, uses that a planner may categorize as "institutional" a citizen may see as having little in common, as their impact on a neighborhood may be radically different. The citizen is thus categorizing uses in other terms than the planner is. An administrator responsible for various penal institutions may have a quite different way to categorize institutions than either the planner or the citizen. There are thus numerous ways to classify different kinds of environments; each may be correct and appropriate in its context, but they may clash when the contexts differ. Though the categorizations of many areas of expertise are often identifiable, those of common perception - of the citizen, as it were - are less self-evident.

Categorization is just as central to the natural environment as to any other area of experience. The management of huge tracts of public lands depends on classification. Both the U.S. Forest Service (1974) and the U.S. Bureau of Land Management (1980) use such classification systems in their management of the public's visual resources. The "character types," "variety classes," and land-form and land-use categories central to these systems do not correspond to the way an untrained citizen would categorize or "see" the environment. Decisions about major changes to the natural environment are based on these expert categorizations.

For each area of expertise, such categorizations take on an air of reality; for the group sharing the code, the categories appear to be reasonable, widely accepted bases for describing the respective environment. It is in the nature of expertise that one not only perceives the world through these categories but no longer remembers that it was ever otherwise. But the basis for environmental categorization for the person-on-the-street, as it were, may not match the land-use, land-cover, forest-practice, or biophysical distinctions.

People react to what they experience in terms of commonalities, in terms of classes or categories. A scene is generally perceived as a particular instance of a larger class of scenes. Let us say we ask people to categorize 40 scenes showing different outdoor environments into groupings they consider important. A park planner may categorize them in terms of activities and their impacts, a botanist in terms of species. The groupings of other individuals may reflect distinctly different aspects of the settings. What might these be?

By examining the preference judgments from a particular study, one can determine some of these commonalities. The procedures for extracting this information – referred to here as Category-Identifying Methodology (CIM) – are discussed in Appendix A.[3] For purposes of the discussion here, it is important to realize that this approach depends on several issues. Clearly, research of this kind requires that there are enough individuals participating in the study. The interest, after all, is in how "people" see the environment and what they prefer. Perhaps less clear is the requirement that one provide enough instances of the kinds of environments one is studying. The issues in sampling the environment are no different than those involved in sampling people. In both cases, one is concerned to determine relatively stable findings that go beyond the particular instance. If we know the preferences of a handful of people, we cannot be confident that these are similar to the preferences of others. If we know preferences for one or two examples of a particular kind of setting, we cannot be confident that these are similar to preferences for other instances of the same category.

Preference ratings thus generate diverse information. One can examine the ratings in terms of how much various scenes are liked or disliked by computing the average rating for each scene. (Results based on such analyses will be discussed in chapter 2.) One can also examine the pattern of the ratings, their relationship to each other. Analyses based on such patterns provide insight into shared categories of perception. Using Category-Identifying Methodology, one can thus extract information about how scenes are grouped. The scenes constituting a category reflect a common perceptual theme; there is no requirement, however, that they be similar in terms of the degree of preference.

Wonderful as such procedures are in terms of identifying thematic commonalities, it is important to remember that these are computational techniques. The outcome of the analysis is the designation that certain scenes "belong together," that they form a group. What it is that they have in common, however, is a matter of interpretation. The researcher (not the computer) provides such an interpretation; others may offer alternative interpretations. For example, the researcher may examine the scenes identified by CIM as belonging to a single category and decide that their common element is the predominance of dense vegetation. Someone else examining these scenes may point out that most of them include a road. The interpretation of the category would likely change with this different emphasis. However, if examination of other groupings derived from the same study reveals roads as predominant in the other groupings as well, then the interpretation based on roads is less germane. Though there is room for difference in opinion when making such interpretations, this is much less a problem than one might expect. Generally, the common themes in a category are easily identified.

Two Examples

Imagine a study where individuals are shown a set of slides that depict various urban settings with which they are not specifically familiar. There are scenes of various contemporary structures that might be used as office buildings, or possibly as residence halls on a campus; there are older buildings that might be used for civic or commercial purposes; there are structures with tall smoke stacks that suggest industrial use, and so on. Some of the buildings are along narrow alleys; some are surrounded by considerable vegetation; some are of unusual architecture, and others are nondescript. As individuals view these slides they rate each scene in terms of their preferences (5 if they like it a great deal, 1 if they don't). Quite aside from determining which scenes are more or less preferred, how would these be categorized? Would scenes with presumably similar functions be grouped together, or would the age of the structure play a determining role? Is the setting, rather than the building, particularly salient?

The answer we obtained in studying this question (Herzog, Kaplan, & Kaplan, 1982*) was that the groupings could not be explained in terms of any one of these distinctions. Using CIM, five categories emerged. By and large these were mixed in terms of function, except that structures that appeared to be factories did not combine with structures that were likely to have nonindustrial use. The age of the building seemed to be a salient aspect of categorization. One of the groupings consisted of older buildings, varying widely in function. Another consisted of more contemporary structures. Scenes that were particularly striking from an architectural standpoint also formed a separate category. The final category, however, was formed on the basis of neither age of the structure nor function. It consisted of the scenes that showed a considerable amount of foliage – an urban nature grouping (Fig. 1.1).

In terms of environmental perception, then, this set of results suggests that the categorization is based on several criteria: function, age, and type of architecture as well as vegetation are reflected in the groupings. For scenes that consist mostly of built elements, distinctions are apparent that relate to architecture and likely use. Settings that are more natural were categorized separately, however.

Let us consider another set of pictures. This time residents were asked to indicate their preferences for each of 40 printed photographs of areas in and near large-scale, low-rise apartment complexes. These areas were similar to settings near where these residents lived and showed a variety of "nature" (e.g., scenes of relatively large open areas surrounded by apartment buildings, smaller natural areas between structures, areas that were neatly mowed, landscaped regions near the buildings, as well as natural areas where buildings were not present in the scene).

How is this type of environment perceived? It would seem reasonable that the underlying commonalities would be on the basis of function – what activities one might carry out. If that were the case, larger open areas where one might play ball would be grouped together by the CIM procedure. Perhaps the architecture of the buildings would be an important consideration, so that scenes of buildings that are similar in style would group together.

The results of this study (R. Kaplan, 1985a*) showed that the size of open space was not a factor in itself; nor was the tidiness or maintenance of the area. Places where one

Figure 1.1 Examples from four of the categories in the Herzog, Kaplan, and Kaplan (1982*) study. *Top left* is representative of Older Buildings category; *top right,* Contemporary Life; *bottom left,* Unusual Architecture; *bottom right,* Urban Nature. For examples of the fifth category, see Figure 2.1.

could take a walk or play ball or sit and sun did not form the basis for grouping scenes. Rather, the results suggested that the basis for grouping was related to two factors: the balance between the buildings and the natural areas and the arrangement of the natural area itself. Scenes in which the apartment buildings (all of these were no more than two stories high) dominated the setting formed a distinct grouping. The fact that there were relatively large open areas and that the grass was neatly mowed did not lead these scenes to be grouped with others which also included open areas but where the buildings were more in balance with the natural setting (Fig. 1.2). At the same time, scenes in which the buildings were not visible at all formed dis-

Figure 1.2 All four scenes depict relatively large open areas; they were not, however, grouped in a single category (R. Kaplan, 1985a*). The scenes in the top row were included in the "faceless" category, and the bottom row scenes were part of a different category.

tinct groupings. There were, in fact, two such groupings (Fig. 1.3). One consisted of a great diversity of settings, with vegetation varying in both density and height. These can be described as "nearby-natural areas." The two scenes that formed a separate grouping both showed large mowed areas with a few trees.

Although kinds of settings represented in this study differed in many respects from the previous example, the results show several interesting consistencies. (1) The categorizations in both studies are not based on any one set of criteria. (2) The issue of "facelessness" (i.e., anonymous and undifferentiated) seems to be an important dimension of categorization. In the second study, the facelessness

Figure 1.3 Though similar in being "natural," with no houses evident, these scenes are nonetheless drawn from two distinct categories (R. Kaplan, 1985a*).

was related to the similarity of the structures and their dominance over the natural elements. In the previous study, the scenes in the factory grouping communicated a "rather desolate feeling by virtue of both their scale and their facelessness" (Herzog et al., 1982*, p. 48). (3) The amount of nature in the scene is an important determinant of categorization.

By exploring the pattern of categories in these two studies, we find that each sheds light on a set of concerns that are intrinsic to the environment it subsumes. At the same time, meaningful comparisons are possible across studies despite their different kinds of settings. If we now expand this comparison to a much greater diversity of environments, spanning perhaps 30 or 40 different studies, we can begin to achieve a greater understanding of how the natural environment is perceived.

TYPES OF PERCEPTUAL CATEGORIES

If we look at the empirically based categories that these many studies have provided, we are faced with an enormous analytic task. There might easily be 150 such categories derived from the CIM results of these studies. Finding a pattern in such a collection can be a challenge. As it turns out, however, the categories can be described in terms of two major types (although some instances may fit both). One of these types can be described as based on *content;* the other focuses on *spatial configuration.*

The content-based categories have as their theme or common characteristic that they deal with specific objects or elements. Contents are relatively easy to name. Waterscapes provide a good example. In fact, "natural environment" itself is such a content domain. As we have seen in the first example, in a study where the scenes are largely of the built environment, "nature" is the basis for a category. Similarly, in studies where the scenes are mostly of natural settings, the nonnatural scenes often form a distinct category (Fig. 1.4). Anderson's (1978*) study provides a further example; one of the categories in his study of forest practices had in common that the scenes all depicted stands of red pine.

The spatial configuration categories are based on the way the elements are arranged in the implied space of the scene. The role of spatial organization has emerged repeatedly in our research as a central theme in understanding people's reactions. We have gradually come to think of people's reactions as an outcome of their imagining themselves in the setting. It is as if one asked oneself what one's possibilities would be if one were there, how easily one could move around in the scene. As individuals rate these scenes they are not, of course, likely to be aware of this process; it appears to occur automatically, effortlessly, and very quickly.

The groupings from the housing study (R. Kaplan, 1985a*) provide examples of both kinds of categories. Recall that two of the categories consisted of scenes showing only natural elements and no residences (Fig. 1.3). One of these, depicting scenes with smooth ground textures and a few trees, could be characterized as Parkland. This category is based on more than a content distinction. The spatial configuration plays a central role here: These settings permit one to see far into the scene without obstruc-

Figure 1.4 These scenes provide examples from four different studies of categories based on human-influenced elements in the context of relatively natural settings. *Top left* is representative of Hudspeth's (1982*) Industry category; *top right*, Hammitt's (1978*) Boardwalk category; *bottom left*, Gallagher's (1977*) Building category; *bottom right*, Kaplan and Herbert's (1988*) Residential category.

tion. By contrast, the scenes of the other natural-areas category consist of a diversity of nearby-natural areas without common qualities with respect to their spatial organization. Here the nature content itself seems to be the basis of the category.

It is perhaps not surprising that content is a major aspect of environmental perception. Grouping what the environment contains into coherent content categories is, after all, basic to cognitive economy. By establishing such groupings it is possible to react to a wide range of different situations in a similar way. Because one is constantly confronting situations that are at least slightly different from any that were previously experienced, the use of categories is not merely convenient but essential if one is to apply one's past learning to new situations. At the same time, however, as will become apparent in the next section, it is by no means obvious what constitutes the specific content domains in the way people experience the environment.

The categories based on spatial configuration, however, provide a different set of insights and challenges. In these cases interpretation of the CIM results may require more than a word or two to describe the grouping. But even when the category can be defined readily (e.g., Parkland), these categories tend to be based on an implicit interpretation of the opportunities and constraints afforded by the space.

It is important to realize that neither type of categorization – whether based on content or spatial configuration – necessarily reflects categories that individuals would provide verbally. The perceived commonalities that lead to these categorizations are based on ratings of scenes; no verbalization is needed. There is no assumption that citizens in Anderson's study would necessarily call the scenes "red pines" (as opposed to pines or firs). Nor are participants likely to label scenes as "impoundments," even though such a category was derived in the storm drain study (R. Kaplan, 1977a*). Note that although the distinction between these two major types of categories is helpful in most cases it must not be assumed that all categories fit clearly into one of these types.

Let us turn now to a more extensive examination of these two types. It is from such an analysis that we can come closer to understanding environmental perception and, more particularly, to seeing how people with no particular expertise in this area experience the natural environment.

Content-Based Categories

A major underlying theme in many of the content-based categories that have resulted from the CIM procedure concerns the balance between human influence and the natural area. Many of the studies show a great sensitivity to kinds and degrees of such balance. In other words, the research results suggest that an underlying component of perception is a classification in terms of the degree of human influence.

As we have already seen, this was the case in the housing complex study (R. Kaplan, 1985a*), where three degrees or levels of balance were identified: the faceless, building-dominated content; open residential areas; and nearby-natural areas with no visible residences. Berris's (1988) study, in a totally different context, showed a similar pattern, with distinct categories showing the faceless, hard spaces of buildings in a setting devoid of plantings as opposed to settings where the natural and built components are balanced. In Gallagher's (1977*) study, the balance between the modern commercial building and the surrounding natural area was also a strong determinant of categorization.

The results of Miller's (1984*) study were particularly instructive in showing an underlying sensitivity to the balance between human influence and the natural environment. Participants in the study were ferry passengers departing from Vancouver, and the scenes they were asked to view depicted places along the British Columbia coastline. Five of the seven CIM-derived groupings in this study reflected different kinds of juxtaposition between the natural and the human-influenced. These ranged from what Miller called Man-Dominated Nature (scenes of intensive development) to Man in Harmony with Nature (scenes with Small Structures in Natural Settings). Miller described the categories between these extremes as reflecting: Man-Intruded Nature (Shoreline Roads), Presence of Man in Nature (indications of logging), and Man's Past Presence in Nature (Clearcuts and Natural Bald Spots).

Quite consistently, in studies of natural settings, where most of the scenes do not include obvious examples of human-sign, those scenes that depict human habitation characteristically emerge as a CIM-based cluster (Fig. 1.4). Herbert's (1981*) study of a rural, largely natural portion of an urbanizing county in Michigan provides a good example. One of the four groupings, Residential, includes virtually all the scenes that show any housing at all. In CIM analyses based on Ellsworth's (1982*) study of rivers and marshes, one of the five clusters consisted of scenes that have in common the presence of human influence. These included a diversity of influences – a bridge, houses, dumped vehicles and trash – suggesting that the perceived human intervention is the likely basis of categorization here.[4]

It would seem then that one of the consistent bases of

perceptual differentiation reflects a "built" as opposed to a "natural" content, when the sampled scenes include both. How broadly the content domain is (perceptually) defined depends on a number of factors. One of these involves the frame of reference provided by the set of pictures to be judged. If relatively few scenes in a particular study reflect clear human intervention, whereas the remainder are more "natural," the few seem to cluster together even if their contents may be fairly disparate. (If such disparate contents were the frame of reference in a particular study, the CIM-based categories would presumably reflect the diversity of content rather than emerging as a single cluster.)

Another factor influencing categorization seems to reflect the apparent importance of the domain. For example, industrial scenes are consistent in forming a category (e.g., Hudspeth, 1982*; Miller 1984*), suggesting that this is a meaningful perceptual category not only for land-use planners and realtors but also for citizens. In other cases, the absence of a category may provide indication that a content domain is not perceived as particularly salient. Roads, although easily labeled and a clear content, have only once appeared as a distinct category across many studies in which they were included. In the S. Kaplan, Kaplan, and Wendt (1972*) study, the Nature scenes included settings with unpaved roads (Fig. 1.5) and even in one instance with a parked car; in Anderson's (1978*) study, scenes with roads appeared in several groupings, suggesting that roads were not perceived as the distinguishing characteristic. In the housing study (R. Kaplan, 1985a*), neither roads nor cars (parking areas) appeared as distinct categories. (Miller's study is the exception here, as the only two scenes with a road, paralleling the shoreline and clearly cutting through an otherwise forested area, did form a category.)

Categorization is also likely to be affected by familiarity or knowledge of the kinds of scenes included in the study. The Red Pine grouping in Anderson's study, for example, would be less likely to appear as a grouping if participants were not local citizens with a great deal of familiarity with such forests. Such familiarity may also account for the lack of grouping of "residential" areas as a common category. In the S. Kaplan et al. (1972*) study, for example, the urban and nature scenes each formed broadly defined categories. By contrast, participants were not identifying residential scenes simply as "residential." This suggests

Figure 1.5 The presence of clear human influence did not prevent this scene and others like it from inclusion in the empirically based Nature category in the S. Kaplan, Kaplan, and Wendt (1972*) study.

that with respect to residential settings finer distinctions are characteristically made. Working-class single-family homes are not perceived as belonging to the same category as architecturally distinctive single-family homes, and apartments distinguish yet another cluster. (In fact, subsequent studies of residential settings [e.g., Frey, 1981, and Widmar, 1983] corroborate this tendency toward much finer discriminations in this context.

To summarize, then, the balance between the built and the natural is a consistently dominant theme in the experience of the environment. Based on empirically derived groupings, human influence emerges as a salient attribute. At the same time, however, results of these studies suggest that human intervention is not a unitary concept. In some instances, a mixture of human influences is perceived as essentially similar, whereas in other cases - mostly in the context of residential land use - human intervention is further distinguished into several distinct categories.

Spatial Configurations

Even more striking than the particular content domains, however, are the categories for which content is not the distinguishing characteristic. In these cases it is the spatial configuration of the scenes that appears to account for the categorization. The "space" in question here is not the two-dimensional space of the picture plane but the inferred three-dimensional space of the scene the photograph depicts.

These categories suggest that an underlying criterion in making a preference judgment is an evaluation of the scene in terms of presumed possibilities for action, as well as potential limitations. Even in the very rapidly made rating, an important consideration involves an assessment of the scene in terms of what it makes possible, what it permits one to do. Gibson's (1979) concept of "affordance" is similar: An affordance is what an environment offers the perceiver, or, in other words, what the perceiver would be able to do in the setting. In the research discussed here the focus is on the assessment of potential actions applied to scenes. People are evaluating an entire setting in terms of potential actions - even though they cannot carry out the actions because the setting is presented photographically.

In virtually all the studies that included a variety of natural scenes, the CIM procedure led to *several* distinct categories that are essentially natural. The existence of separate categories within one study suggests that the presence of vegetation (as a content domain) is not the only distinguishing characteristic; other properties of the scenes must be salient in forming distinct groupings. Comparison of these categories across studies suggests two themes as particularly prominent in distinguishing among such separate nature groupings. One of these entails the *degree of openness,* and the other involves *spatial definition.*

The wide-open theme has been evident in the CIM results of many studies. The scenes comprising a Wide Open category generally lack any particular differentiating characteristics, and the sky occupies a considerable portion of the scene (Fig. 1.6). Such landscapes might be of farmland (R. Kaplan, 1977a*) or of unused roadside land (Ulrich, 1974*) or of the Low Flat Shorelines (Miller, 1984*). They might also be scenes of Marshes (Ellsworth, 1982*) or the Bog mat (Hammitt, 1978*).

In Anderson's (1978*) study two separate clusters can be included under the wide-open heading. Participants in this study were local people, knowledgeable about forest prac-

Figure 1.6 Examples of categories reflecting wide-open, undifferentiated spatial arrangements (from scenic roadside study, R. Kaplan, 1977a*).

tices (and dependent on forestry as the major factor in the local economy). The CIM-based results thus provide an indication of their perceptions of the local landscape. One of the two wide-open categories Anderson called Heavily Manipulated Landscapes. It included scenes of recent clearcuts, cutover stands, cleanup after harvest, and poorly stocked areas. The other category, Open, Unused Land, Anderson says "may be considered as wildlife openings, old uncultivated fields, or meadows." The fact that two separate categories emerged poses an interesting situation.

Based on a visual analysis of the scenes, it is true that both categories are generally open and lack distinctive features. The two categories differ, however, in the land uses they reflect or, more specifically, in the degree of human influence they represent. Thus, although the openness of the scenes is an indication of the role of spatial configuration, the emergence of two separate categories of this kind suggests that content also plays an important role in the perceptual process.

Scenes that lack openness also characteristically emerge as a separate category. Here too there is a lack of focus or of any differentiating characteristics, but rather than giving a sense of endless open space, the view is blocked. The inability to see into the scene may be created by dense foliage, by an embankment, or by other major obstructions. From an affordance point of view, either end of the openness continuum is limited. In a blocked setting, one's ability to see or move would be seriously limited. In a wide-open situation, by contrast, movement and visual access may be unlimited, but finding one's way back may be made difficult by the lack of distinguishing features.

In the second major type of spatial configuration categories these problems are considerably reduced. In these instances the scenes can be characterized as having *spatial definition.* Space can be defined in many different ways. What they have in common involves the presence of distinct edges or landmarks that help structure the setting. Examination of the clusters that have appeared in the various studies we are considering here is helpful in understanding this notion.

Consistently, across many studies, scenes that are relatively open but have some distinct trees form a separate grouping. It is difficult to judge the depth of an undifferentiated surface; even a relatively small number of trees, however, contributes markedly to the spatial definition of a scene. In most studies an appropriate name of such relatively open yet spatially defined clusters might be *parklike* (Fig. 1.7). As already noted, this appeared in the apartment complex study (R. Kaplan, 1985a*); similar categories were also found in the study of the storm drain (R. Kaplan, 1977a*), Ulrich's (1974*) study of roadside scenes, Hudspeth's (1982*) study of waterfront revitalization, and Herbert's (1981*) cluster of Manicured Landscapes in an urbanizing rural setting. In the context of less

Figure 1.7 Examples from categories that are relatively open, yet show clear spatial differentiation.

managed natural settings, Woodcock's (1982*) Savanna scenes also illustrate this phenomenon.

The comparison of these studies shows interesting parallels for environments that are relatively open and for those that are generally forested. In the relatively open natural settings, two major types of categories that are based on spatial configuration appear quite consistently: the wide-open areas with little differentiation and the relatively open areas with clear spatial definition. In forested environments, comparable themes appear: Separate categories emerge for forests that are dense and seemingly impenetrable and for relatively open forests. In the forest, then, the dense, blocked configurations are parallel to the wide-open area. Both are characterized by the degree of openness. And both provide less spatial definition than do their better-structured counterparts.

Figure 1.8 Forest-based categories that differ in whether they are relatively open (*top row*) or more blocked (*bottom row*).

Several studies provide examples of these two themes in the forest context (Fig. 1.8): In Woodcock's (1982*) study, the rain forests tended to be grouped together and constituted the more "blocked" landscapes, whereas the mixed hardwood forests were generally more open. In both the scenic route study (R. Kaplan, 1977a*) and a study of wilderness areas in Michigan's Upper Peninsula (R. Kaplan, 1984b*), the degree of openness of the forest was a key basis for categorization. In Anderson's (1978*) study, the grouping that exemplified forest scenes with spatial definition was the Planned Spacious Openings and Scenic Roads grouping. Here spatial definition is enhanced by the openings in the forest and, in some instances, by paths or roads. Generally, in the forest context, lack of underbrush and sufficient openness that one can see through and among the trees enhance the spatial definition.

Thus, both openness and spatial definition constitute salient perceptual categories that emerge from the empirical results of preference judgments. They suggest that perception entails a very rapid (albeit unconscious) assessment of what it is possible to do in the setting. In the dense, blocked forest views, where there is considerable understory or a mass of foliage, neither visual nor locomotor accessibility is apparent. The opportunities for acquiring knowledge are seriously impaired. In the relatively more open forests and the more open areas with greater spatial definition, by contrast, there is a sense that one could function more effectively either because the transparency among the trees permits increased visual access or because the smoother ground texture suggests that locomotion could be accomplished relatively easily. The presence of landmarks, of features that contribute spatial distinctiveness such as a few trees in an otherwise relatively open setting, provides spatial definition and suggests that wayfinding will be possible. On the other hand, scenes reflecting a sameness – either in a wide-open area lacking structure or in a dense woods – suggest a greater likelihood that one might get lost.

SOME CONCLUDING COMMENTS

There are many ways to categorize a particular environment. Use of preference ratings by untrained participants (in conjunction with a Category-Identifying Methodology) yields categorizations that are distinctly different from those generated by various professionals. The groupings identified across a great diversity of studies provide some insights into the salient aspects of environmental perception.

The use or function of the land is part of the implicit categorization underlying environmental perception. Thus certain kinds of uses tend to emerge as distinct clusters; industrial land uses serve as a good example. But land use is also too broad a category in many cases; the empirically derived groupings indicate that neither residential land uses nor natural areas tend to be experienced as unitary content domains. Rather, these are disaggregated into separate groupings reflecting subtler differentiations. In fact, the balance between the built and the natural is an important basis for such differentiation.

Content is only part of the basis for the way the environment is perceived. People are extremely sensitive to the spatial properties of the environment. It is apparent that in the rapidly made and largely unconscious decision regarding preference, there is an assessment of the glimpsed space and its qualities. This rapid assessment appears to be heavily influenced by the potential for functioning in the setting. Thus indications of the possibility of entering the setting, of acquiring information, and of maintaining one's orientation emerge as consistently vital attributes. The consistent presence or absence of these attributes frequently provides a basis for the formation of a coherent perceptual category. Wide-open, undifferentiated vistas and dense, impenetrable forests both fail to provide information about one's whereabouts, and both consistently appear as distinct perceptual categories. On the other hand, scenes that convey a sense of orderliness (such as parklike and manicured settings) also tend to form a distinct category. In such settings, the smooth ground texture affords prediction about how one could function in the setting. Similarly, forests that are more transparent, with light filtered through the trees and with suggestions of paths, provide information about accessibility and direction. Thus a major underlying basis implicit in these categorizations is an assessment of the possibility of functioning in the setting.

It is striking how profoundly information-based these categorizations have been. The majority of the expert-generated category systems have little to do with the way people process information. Results of the various studies summarized here suggest that what people experience as salient in the landscape involves informational patterns, readily interpretable in terms of requirements for adaptive behavior. The way space is organized provides information about what one might be able to do in that space. A relatively brief glance at a scene communicates whether there is room to roam, whether one's path is clear or blocked.

Thus there appear to be both an empirical and a theoretical basis for categorizing landscape scenes. As is often the case with a satisfying research experience, these categories would have been hard to anticipate but in retrospect make intuitive as well as theoretical sense.

NOTES

1. Levin's (1977*) study was an early effort that showed strong similarities in reactions to on-site and photographic representations of the same environments. Shuttleworth (1980) and Ulrich (1979) discuss some of the distinct advantages of using photographs. Zube, Simcox, and Law (1987) provide a review of some of this literature and also argue for its validity and reliability. Kellomaki and Savolainen's (1984) study, carried out in Finland, is noteworthy because they used both "ordinary city dwellers" and students trained in forestry to study the differences between field and laboratory ratings of forest landscapes. Coeterier (1983) reports that photographs can be substituted for actual settings more reliably for "small-scale landscapes" than for "large-scale landscapes with microrelief."
2. Categories and related concepts (e.g., prototypes, schemata, natural categories) play a central role in modern cognitive psychology. The findings are generally consistent with the approach we have taken, although much of the work in the area (e.g. Rosch, 1978; Tversky & Hemenway, 1983) has been (from our perspective) hampered by the traditional information-processing emphasis on alphanumeric material as opposed to drawings and photographs. The most recent trend in this area is away from the strict logical definition of a category and toward concepts with fuzzier boundaries. Some researchers feel that even the idea of fuzzy boundaries is insufficient to account for the subtlety of category use and argue instead for the centrality of some sort of implicit theory a person calls on to decide whether something does or does not belong in a category. For stimulating discussions of recent thinking about categories, see Lakoff (1987); Murphy and Medin (1985); and Posner (1986).
3. CIMs previously were designated *content*-identifying methodologies in S. Kaplan (1979a). The abbreviation has not changed, but *category* is adopted here as a more descriptive name for the procedure. Since some categories are based on spatial configuration rather than on content, this confusion is avoided.
4. Though these scenes received relatively low preference ratings, it was not the distaste for the intervention that was the basis for the perceptual grouping. The marsh scenes were no better liked but were a clearly distinct theme. In other words, the categorization is based on a similarity in reaction to a group of scenes, not necessarily on whether they are liked or disliked. Signs of blatant human influence in an otherwise natural setting are perceived as a distinct category.

CHAPTER 2

THE PREDICTION OF PREFERENCE

THERE is no question that people's tastes differ. With respect to culinary tastes, the diversity is evident around the dinner table at home or in a restaurant. With respect to clothes, differences in preference are also readily apparent. Even with respect to natural environments, the variation in preference is quickly evident. Consider pictures and posters of natural places that people display on their walls. There are the grand, the intimate, the awe-inspiring, the tranquil, the ephemeral, the commonplace, the wild, the tame, the representational, and the abstract. A look around a custom frame shop provides a sampling of this diversity.

It is easy, however, to draw false conclusions from such diversity. One frequent generalization is that "there is no accounting for people's tastes" or that "beauty is in the eye of the beholder." These statements suggest that taste is random, as variable as people are. Another false inference is that tastes are frivolous or whimsical or that the issue of preference is simply not of great moment. It might be nice to have a vase full of fresh flowers, but one could certainly survive without.

If indeed tastes were random and the question of preference were of no great consequence, there would be little reason for this chapter. The study of environmental preference, however, shows remarkable consistency, despite demographic differences and across diverse settings. Given these results, there is reason to suspect that environmental preferences provide a glimpse into some essential ingredients of human functioning. The purpose of the present chapter, then, is to explore the question of preference for natural environments in a relatively generic sense – in terms of attributes that are likely to be pertinent across the diversity of humankind. (In chapter 3 we will return to the question of differences in preference.)

PERCEPTION AND PREFERENCE

The discussion of perceptual categorization in chapter 1 was based on ratings of scenes in terms of how much the viewer liked them – in other words, preferences. As we described, Category-Identifying Methodology (CIM) provides an approach to analyzing these ratings by extracting common patterns in the way people respond. By comparing the categories that emerged in numerous studies we found that these fall into two major types: content-based and spatial configurations. In chapter 1 these types of categories were examined as a way to understand how the environment is experienced. Preference ratings were used as a vehicle to understand perception – but the preferences per se were not included in the discussion.

Perception is quite obviously important to survival. It is, for example, clearly adaptive to be able to perceive danger. But being able to perceive what is safe and what is dangerous is not enough. If the information an organism acquires through the power of perception is to aid in its survival, it is essential that it not only perceive what is safe but also prefer it (Appleton, 1975; S. Kaplan & Kaplan, 1982). Such a pattern of preferring a suitable habitat over an unsuitable one is widespread in the animal kingdom and may well be characteristic of humans as well.[1] If this were the case, one would expect that what is basic to perception must also be important to preference.

It is time, then, to look at the perceptual categories to see what they reveal about preference. To do this we will use the same studies that were the basis of the discussion in the preceding chapter (identified by * in the discussion here and summarized in Appendix B). Recall that participants in a particular study are asked to view a series of scenes (photographs or slides) and to indicate their preference for each one, using a 5-point rating scale. These preference ratings are analyzed using CIM procedures (see Appendix A), resulting in a number of categories that reflect commonalities in rating patterns. The categories tell us how the environment that is presented in the scenes is perceived by the participants. To determine how much each of these perceptual categories is liked, the preference ratings for the scenes comprising each category are averaged. These mean preference values are the basis of the analysis here.

Before examining these preferences in terms of the different types of categories, it is useful to provide some sense for the mean preference values that are characteristic in such studies. The groupings typically consist of anywhere from 3 to 12 scenes. For an average preference rating of a grouping to be near the extremes of the 5-point scale would require great unanimity. This is unlikely not only because people vary but also because the various scenes that comprise a perceptual category will differ in details that can affect the degree of preference. In fact, it is most unusual to find a grouping with a mean of 4.0 or higher or one that is at 2.0 or lower. In general, means below 3.0 reflect "low" preference, and those at 3.7 and above can be considered relatively "high."

In addition to considering these absolute levels, it is important to consider the ratings relative to other category means within a given study. Given the very diverse contexts for the studies, the range of ratings differ somewhat. For example, in the study of the nearby storm drain system (R. Kaplan, 1977a*), ratings tended to be relatively low in general - perhaps because most of the scenes depicted near-home places that needed some care. The most preferred scene in the study was from another area entirely, unknown to the participants and extremely well maintained. By contrast, in Hammitt's (1978*) study, ratings tended to be relatively high. In this situation, people were visitors to a featured attraction (a bog trail at Cranberry Glades in West Virginia), and most of the scenes were taken in that setting. Here the scenes taken elsewhere generally received lower ratings.

Content-Based Categories

As discussed in chapter 1, the content-based categories that were empirically derived across the various studies reflected a concern for the balance between human influence and natural areas. In studies that included industrial scenes these invariably emerged as a category. Not surprisingly, these categories were also consistently low in preference. In both Hudspeth's (1982*) and Miller's (1984*) studies, for example, the industrial groupings received by far the lowest preferences (mean of 1.8 and 2.1, respectively). The Alley/Factory grouping in the Herzog et al. (1982*) study, with a mean of 1.6, included a mixture of industrial and other barren-seeming settings (Fig. 2.1).

It is not only the industrial scenes that receive low preference ratings. In the vast majority of studies, the categories that reflected the clearest human influence were relatively lowest in preference. The human influences varied considerably in type. In the S. Kaplan et al. (1972*) study, it was the Urban that formed a single category (mean 2.2). In the multiple-family housing study (R. Kaplan, 1985a*), the Building-Dominated group (mean 2.2) (shown in Fig. 1.2, top row) reflected the strongest sense of human influence. Herbert's (1981*) study included a rural Residential category; in this case the buildings were far less dominant in the landscape, but relative to the other

Figure 2.1 Two scenes from the Alley/Factory category in Herzog, Kaplan, and Kaplan (1982*) study.

scenes in the study they showed the strongest human influence, reflected in a mean rating far lower than any others in the study (2.8). The Human Influence category based on Ellsworth's (1982*) data showed a great mixture of intrusions, but the mean rating was similarly low (2.5).

Anderson's (1978*) study is striking for having a strongly unpreferred category (mean 2.3) where the human influence involves no built elements. Rather, these scenes show the results of clearcutting; Anderson refers to the category as Heavily Manipulated Landscapes. Given that the viewers in this case were local residents who are highly familiar with such scenes and whose economy depended on the lumber industry, the finding is even more noteworthy.

Not all the human-influence content categories, however, were low in preference. Hudspeth's (1982*) Boats category received midrange ratings (3.6), and Miller's (1984*) Small Structures in Natural Settings category was second highest among the categories in his study at 3.9. Hammitt's (1978*) Boardwalk category (included in Fig. 1.4) was high in preference (mean 4.1). Here a wooden boardwalk was the common theme in four scenes that were otherwise totally natural. In both the Miller and Hammitt instances, the human influence is central to the content, but the built component is in keeping with the setting and does not dominate the natural elements in the scenes. Figure 2.2 provides a summary of the relative preferences of these categories, based on 10 studies.

At the other end of the continuum, the most preferred content-based categories have generally been ones where nature is dominant in the scenes. The two studies used as examples in chapter 1 provide good examples: In the multiple-family study (R. Kaplan, 1985a*) the Nature category received a mean of 4.0. In the Herzog et al. (1982*) case, the Urban Nature group was by far the most preferred (3.6, two scale points higher than the Alley/Factory grouping). In many other studies the means have ranged between 3.5 and 3.7 – always significantly higher than for the categories that were relatively less natural within the same study. In analyses based on Ellsworth's (1982*) data, for example, though all scenes showed waterscapes, the category with snow-peaked mountains in the background was a full scale point higher in preference than the one with human intrusions.

4.2
4.1 Boardwalk (Hammitt, 1978*)
4.0
3.9 Small structures in nature setting (Miller, 1984*)
3.8
3.7
3.6 Boats (Hudspeth, 1982*)
3.5
3.4
3.3
3.2
3.1
3.0
2.9 Building (Gallagher, 1977*)
2.8 Rural residential (Herbert, 1981*)
2.7
2.6
2.5 Human influence (Ellsworth, 1982*)
2.4
2.3 Heavily manipulated (Anderson, 1978*)
2.2 Urban (S. Kaplan et al., 1972*); Building-dominated (R. Kaplan, 1985a*)
2.1 Industrial (Miller, 1984*)
2.0
1.9
1.8 Industrial (Hudspeth, 1982*)
1.7
1.6 Alley / Factory(Herzog et al., 1982*)
1.5

Figure 2.2 Relative preferences for categories reflecting human influence in context of scenes that are mostly natural.

Spatial Configurations

In chapter 1 we described the second type of perceptual categories as reflecting the interpretation the observer intuitively and unconsciously makes about the ability to function in the pictured "space." These interpretations are based on how the space is organized. Recall from that discussion that two components are particularly salient in these perceptual categories: degree of openness and spatial definition. We suggested that the preference judgments seem to reflect a very rapid assessment of what one could do were one in the scene. The perceived openness (either wide-open or seemingly blocked) seems to be critical to these rapid assessments. Similarly, a sense of depth or focus affects the judgment about how readily one could function in the pictured setting.

One of the most consistent categories to emerge across the various studies involves areas that are very open, with the sky taking up a large portion of the scene and a lack of distinctive features in the foreground (Fig. 1.6). These

Figure 2.3 Examples from the two categories in Anderson's (1978*) study that reflect wide-open configurations. Scenes in the top row are taken from the Open, Unused Land category, and the bottom row scenes are from the Heavily Manipulated Landscapes category.

groupings are generally among the lowest in preference within any particular study, with preferences ranging from 2.3, in the case of the Marsh grouping based on Ellsworth's (1982*) data, to 3.1 in Herbert's (1981*) Pastoral grouping. Scenes of distant farmland, both in the scenic highway (R. Kaplan, 1977a*) and in the R. Kaplan, Kaplan, and Brown (in press*) studies, were similarly low in preference, with means of 2.7 and 2.9, respectively. The Bogmat scenes in Hammitt's (1978*) study are exceptional for their considerably higher preferences (mean 3.7), although this category was distinctly less preferred than the other categories to emerge in that study.

Recall from the previous chapter that in Anderson's (1978*) study two categories emerged that fall under this "wide-open" heading. Here spatial configuration and content considerations both play a role, as the two categories are distinguishable in terms of the degree of human intrusion (Fig. 2.3). As already mentioned, timber harvesting is an essential aspect of the local economy for participants in Anderson's study, and yet these residents far preferred the scenes of Open, Unused Land to those depicting the outcomes of relatively recent harvesting (means 3.2 and 2.2, respectively).

Though extreme openness is not a preferred characteristic, the other extreme fares no better. Such categories consist of blocked views or very dense vegetation (included in Fig. 1.8), suggesting that both visual and locomotor access would be difficult. Woodcock's (1982*) study included two such categories: Dense Hardwood Forest and Rain Forest (means 3.0 and 2.8, respectively). Gallagher's (1977*) prairie restoration study also yielded a "blocked" grouping, consisting of the tall grasses (mean 2.5) (Fig. 2.4).

Figure 2.4 Examples of blocked views taken from Gallagher's (1977*) study. These are scenes of the Prairie planting.

In terms of preferences, the spatial configuration that consistently generates favorable responses involves areas that are open, yet defined. Characteristically, these have relatively smooth ground texture and trees that help define the depth of the scene (Fig. 1.7). Such categories can be called *parklike* or *woodlawn* or *savanna*. Mean preferences for these are always among the highest, ranging between 3.7 and 4.2 (e.g., Wookcock's [1982*] Savanna, Anderson's [1978*] Planned Spacious Openings, Gallagher's [1977*] Complex, R. Kaplan's [1985a*] Parkland, Herbert's [1981*] Manicured Landscapes, and the Woodlawn grouping in R. Kaplan et al. [in press*]. Ulrich's (1974*) study also included a clear example of such a category, but because he used a 6-point scale the preference ratings are noncomparable; nonetheless, the preferences for the parklike scenes were by far the highest. Figure 2.5 provides a summary view of these relative preferences, based on 10 studies. (See also Ulrich, 1986, and Schroeder and Green, 1985, for further studies showing the strength of preference for such configurations.)

4.4
4.3 Open forest (R. Kaplan, 1984*)
4.2 Parkland (R. Kaplan, 1985a*)
4.1
4.0 Complex (Gallagher, 1977*)
3.9 Manicured (Herbert, 1981*)
3.8 Planned spaciousness (Anderson, 1978*); Parklike savanna (Woodcock, 1982*)
3.7 Woodlawn (R. Kaplan et al., in press); Bogmat (Hamitt, 1978*)
3.6
3.5
3.4
3.3
3.2 Open, unused (Anderson, 1978*)
3.1 Pastoral (Herbert, 1981*)
3.0 Dense forest (Woodcock, 1982*)
2.9
2.8 Rain forest (Woodcock, 1982*)
2.7 Distant farmland (R. Kaplan, 1977a*)
2.6
2.5 Prairie (Gallagher, 1977*)
2.4
2.3 Marsh (Ellsworth, 1982*)
2.2 Heavily manipulated (Anderson, 1978*)
2.1
2.0

Figure 2.5 Relative preferences for categories reflecting spatial configuration.

Summary

The perceptual categories are important for identifying characteristics that are salient in the way the environment is experienced. At the same time, they also provide insight into patterns that are liked and disliked. In general, the content-based categories suggest that in the context of natural environments those with human intrusions are less preferred and those where nature dominates the built elements receive much more favorable response. In other words, people are sensitive to the way human influence is positioned in the context of the natural setting. All too often these human influences are intrusive. But human influence can be incorporated in a way that is responsive to the form and characteristics of the natural features.

The categories based on spatial configurations suggest that the perceptual process is sensitive to the relative openness and spatial definition of the setting. Preferences for the resulting categories show that preferred settings are those where it is easiest to extract the information needed to function. In very open and in blocked areas it is more difficult to anticipate what might happen, and preferences tend to be relatively low. In spatially defined areas and in open forests, by contrast, it is far easier to judge where one can venture safely and what to expect. Such categories tend to be highly favored.

THE PREFERENCE MATRIX

The empirically derived perceptual categories provide one way to examine environmental preferences. Let us turn now to a different way of looking at the preference ratings. This approach is by far the most straightforward, low-tech way to explore research results. Rather than looking at commonalities in patterns of ratings (and relying on CIM "technology"), this procedure simply examines the mean preference rating for each scene.

Using the same set of studies (all of which included 5-point preference ratings), an average rating is computed for each scene in a particular study. One then selects the most and least preferred scenes for closer examination. (The decision as to what constitutes "most" and "least" can be based on some arbitrary cutoff such as "means greater than 4.0 and less than 2.0" or the top and bottom six to eight scenes.) When spreading these scenes out before one, in many cases the interpretation of preference is readily apparent.

Using this procedure, as with the CIM-based analyses, the content of the scenes often appears as the underlying basis for the different preferences. Waterscapes are often among the highly preferred settings (except, of course, in studies that consist mostly of waterscapes, such as the storm drain study [R. Kaplan, 1977a*] and Ellsworth's [1982*] rivers and marshes study). Since industrial scenes are so consistently down-rated, they are often in the least preferred pile.

Here again "content" is not a sufficient basis for understanding the preferences. When looking at the most and least preferred scenes one often finds some other characteristics that they have in common, although these may be more difficult to express. It is such "eyeballing" that informs intuitions. By looking at highly preferred and unpreferred scenes we came up with some hunches about preferences. These hunches inspired the next studies, which led to revised intuitions. Gradually a framework developed that we have called the Preference Matrix. With further research and with efforts to use the framework, it has continued to evolve.[2] We cannot promise that the process is yet complete. Let us turn, then, to its current rendition.

Preference and Human Needs

As we said earlier, human functioning depends on information. People seem to be extremely facile in their ability to extract information from the environment. Even the very briefest glimpse of the passing landscape provides information. This information does not depend on posted signs or neon lights. It is far subtler and generally not a part of one's awareness.

The Preference Matrix is divided into two domains representing two critical facets of people's relationship to information. We will first consider these two major domains, each of which has two parts, and then look at the features of the environment that enhance each of the four cells in the matrix.

The first domain involves two major categories of human needs: understanding and exploration. The need to *understand,* to make sense of what is going on, is far-reaching in its expression. Even reasonable, kindly people can become hostile and angry when they cannot comprehend material that seems to be necessary to functioning. In fact, there are times when such frustration is readily apparent even when the failure to understand would seem to have little immediate consequence. Occasionally one can witness such reaction when individuals view "modern" art that fails to make sense to them. The hostility evident at many public meetings may also be a reflection of the citizens' inability to understand the technical material thrust upon them.[3]

Understanding the environment, like anything else, is at least partially dependent on prior experience. Many animal signs, such as nests or lodges or hives, may make little sense to one individual and yet be readily recognized or interpreted by someone with experience. Similarly, the environmental patterns associated with different forest practices may be interpreted very differently depending upon one's ability to understand them. Buhyoff, Leuschner, and Wellman (1979) have shown that preferences for trees that are diseased (orange-brown foliage from southern pine beetle damage) are higher when people are not knowledgeable about the cause of the coloration. Similarly, Americans looking at Australian landscapes are not likely to understand that trees without leaves are diseased rather than showing seasonal variation, because they are not aware that Australia has no native deciduous trees (R. Kaplan & Herbert, 1987*).

Despite differences in background knowledge, however, the need to understand is quite general, and certain environmental attributes play a strong role in making such comprehension more likely. If humans have a strong need to understand, it seems reasonable that preferences would be greater when comprehension is facilitated. We will look at some of these attributes shortly.

Though understanding is important, it is often not sufficient. People also prefer circumstances that require them to expand their horizons, or at least circumstances where such enrichment is a possibility. The second category of human needs, then, is the need to *explore,* to find out more about what is going on in one's surroundings. Exploration is an important element in accumulating experience. It inclines one to expand one's knowledge as well as to increase one's capacity to understand previously confusing situations. At the same time, it provides a way to deepen one's grasp, by inquiring into new facets of a familiar situation.

Exploration, like understanding, is also greatly affected by previous experiences. As is true with understanding, however, even in the absence of prior knowledge, the likelihood that one will venture forth and seek more information can be strongly affected by environmental attributes. If the need to explore is a pervasive human need, it is also reasonable that preferences would be greater where exploration is facilitated.

The second domain of the Preference Matrix involves the *degree of inference* that is required in extracting the needed information. It may be easiest to think of this factor in terms of the two-dimensional and three-dimensional aspects of the visual environment.

The two-dimensional aspects, or the picture plane before one, involve information that is immediately available. It is perhaps easiest to think of this in terms of a photograph of any given landscape. Consider the patches of light and dark areas distributed across the paper on which the photograph is printed. This pattern or organization of light and dark on the photograph constitutes the basis of this level of analysis. Processing information at this level requires very little inference.

The three-dimensional pattern of the actual or depicted space requires greater inference. When we consider the depth of the scene, the space unfolds before us. A setting with partially obscured elements, or with features that mark the distance available to the viewer, provides information that requires more interpretation. As we saw in chapter 1, the three-dimensional pattern invites one to imagine oneself in the scene. Thus there is greater emphasis on what might be seen from a different vantage point as opposed to what is immediately apparent.

Informational Factors in Preference

The Preference Matrix, up to this point, has been concerned with the two basic informational needs – understanding and exploration – and with a dimension that considers how readily available the information is. The combination of these two domains yields four distinct combinations, or patterns (Table 2.1).

Table 2.1. *The Preference Matrix*

	Understanding	Exploration
Immediate	Coherence	Complexity
Inferred, predicted	Legibility	Mystery

These combinations, or informational factors, are given names here that have been part of the landscape assessment literature. It is important to point out, however, that these factors have been defined in many ways. Our usage is strongly affected by the context of the matrix. In other words, we see these factors in terms of the way they help one understand or explore, and in terms of a two- or three-dimensional analysis of the scene. Let us turn, then, to an examination of the environmental attributes or the characteristics of the way the environmental scene is organized that help define each of these four combinations.

Complexity. Two very different disciplinary traditions have focused on this informational factor. Psychologists interested in the area of aesthetics for many years concentrated on the role of complexity. Using artificially constructed stimulus patterns, the basic finding of this "experimental aesthetics" work was that people prefer patterns that are at neither the high nor the low end of the complexity continuum (Day, 1967; Vitz, 1966).

The other tradition that has focused on complexity differs in two important respects. The emphasis has been on the environment, as opposed to artificial stimuli, and the work has not been based on research. Many of the visual resource management systems that federal land agencies have adopted (e.g., U.S. Bureau of Land Management, 1980; U.S. Forest Service, 1974) have assumed that diversity is an important component of scenic quality; settings with greater diversity receive higher ratings in these scoring systems.

In the studies under consideration here, Complexity is defined in terms of the number of different visual elements in a scene; how intricate the scene is; its richness. It thus reflects how much is going on in a particular scene, how much there is to look at – issues that call upon the picture plane, as opposed to depth cues. Clearly, exploration is en-

Table 2.2. *Relationship between Coherence and Complexity*

	Complexity	
	Low	High
Coherence		
Low	Not much there	Visually messy
High	Clear and simple (boring)	Rich and organized

hanced when there is more variety in the scene, when there is the suggestion that there are more different things available. It could be argued that Complexity provides content or things to think about.

Coherence. This informational factor has received relatively little study. Coherence, in our perspective, helps in providing a sense of order and in directing attention. A coherent scene is orderly; it hangs together. Coherence is enhanced by anything that helps organize the patterns of brightness, size, and texture in the scene into a few major units. Such features as repeated elements and uniformity of texture, both examples of redundancy, help to delineate a region or area of the picture plane. Properties of a scene such as texture as well as the size and location of the various uniform areas are assumed to be the province of the "location system" (S. Kaplan, 1970; S. Kaplan & Kaplan, 1982), an ancient neural structure that processes visual information with great speed and little need for inference.

As is true with Complexity, Coherence involves relatively little inference, relying on the two-dimensional aspect of the setting. A coherent setting contributes to one's ability to make sense of the environment. An orderly situation is easier to understand; it may not be more inviting of exploration. The trade-offs between Coherence and Complexity are interesting to consider. One is tempted to consider a messy setting as overly complex. More likely, given this framework, it lacks in Coherence. It is important to realize, however, that a scene can be high in Complexity and in Coherence at the same time (Table 2.2).

Legibility. Kevin Lynch introduced this concept in his profound little book, *The image of the city:*

This book will consider the visual quality of the American city by studying the mental image of that city which is held by its citizens. It will concentrate especially on one particular visual quality: the apparent clarity or "legibility" of the cityscape. By this we mean the ease with which its parts can be recognized and can be organized into a coherent pattern. (Lynch, 1960, pp. 2 – 3)

Lynch's notion of legibility includes all of what we are calling understanding, since he points to the necessity of both coherence and structure. In his discussion of legibility, however, he showed the centrality of orientation and way-finding to the ability to build a mental map of the setting. We have adopted Lynch's term to refer to this more structural (and inferential) aspect of understanding.

As we define it, a *legible space* is one that is easy to understand and to remember. It is a well-structured space with distinctive elements, so that it is easy both to find one's way within the scene and to find one's way back to the starting point. It is also important that the objects be identifiable (e.g., Herzog, 1984*, 1985*, 1987*) and the scene be experienced as interpretable. (The concept of "fittingness," as used by Wohlwill and Harris, 1980, may similarly reflect the roles of identifiability and interpretability.) Legibility thus entails a promise, or prediction, of the capacity both to comprehend and to function effectively.

Landmarks and regions (concepts for which Lynch provided perceptive analysis) that are distinctive and meaningful are important aids in achieving legibility. These help in building a cognitive map, in the "memorability" of a scene (or city, for Lynch). Such distinctive elements or regions are particularly useful in giving one the sense that one could both comprehend and move effectively within the setting.

Mystery. This informational factor also involves promise, but here it is the promise that one could learn more. Something in the setting draws one in, encourages one to enter and to venture forth, thus providing an opportunity to learn something that is not immediately apparent from the original vantage point. There are several ways that scenes or settings can suggest that there is more information available. Some classic examples include the bend in the path and a brightly lit area that is partially obscured by foreground vegetation. Partial obstruction, often from foliage, and even modest land-form changes can enhance this sense of Mystery.

Thus, for Mystery (as defined here) to be present, there must be a promise of further information if one could walk deeper into the scene. This necessarily implies that it would be possible to enter the scene, that there would be somewhere to go. It is important to contrast Mystery with surprise. A path leading to a visible closed door suggests surprise but not Mystery. For the latter, the change in vantage point needs to provide information that is continuous with what is already available, rather than a surprise. Given the continuity, one can usually think of several alternative hypotheses as to what one might discover – in other words, there is both inference and a sense of exploration.

Designers have long recognized the powerful effect of these qualities. Hubbard and Kimball used the word *mystery* in their 1917 text, *An introduction to the study of landscape design*. Their use of it extended beyond the restricted sense intended here. For example, their example of mystery, when "the sheer multiplicity of detail prevents our clear comprehension of the landscape, as when we look at the misty leaves and branches of a thick deciduous wood in early spring," would probably be a better example of Complexity as used here. However, they provide many useful examples of situations that suggest Mystery. For example, they write of the "inability to see the landscape with any distinctness, as for instance when the scene is shrouded in haze or in a snow storm or in darkness." Another of their examples – "the foreground is clearly seen, but . . . an important part of the landscape known to be present is nevertheless concealed, as where a river or a road winds out of sight behind some intervening barrier" – corresponds closely to Mystery as we define it. Thus we would agree with them that "the effect of mystery is the result of impossibility of complete perception," but not all instances of incomplete perception qualify for Mystery. The strong role Mystery has played in preference supports their assertion that "it is a pleasant challenge to the imagination which sets the observer to trying to determine for himself by closer investigation what is concealed from his first glance, or if this be impossible, to filling in and completing the unseen landscape according to the play of his own fancy" (Hubbard & Kimball, 1917, p. 82).

Simonds (1920) also wrote of mystery, which he equated with curiosity:

> The interest in any view is increased by an arrangement that piques one's curiosity. In illustration of this, think of woods into which one gets glimpses leading to unknown depths, bays of lakes disappearing behind islands or promontories, lawns partly hidden by projecting groups of shrubs. These give possible opportunities for making discoveries, and such opportunities compete with variety in giving spice to life. The shape of a tree, the graceful or strong arrangement of its branches, the outlines and texture of its leaves, the color and forms of flowers, the curves of the earth's surface, the reflections in water – are all objects of interest and beauty, but beyond all these in making a view interesting are the elements of curiosity and mystery. (Pp. 13–15)

Cullen's (1961) use of *mystery* – "where anything could happen or exist, the noble or the sordid, genius or lunacy" (p. 51) – does not match ours. However, his discussion of "those aspects of here and there in which the here is known but the beyond is unknown" does partially overlap with our concept. In particular, his use of *anticipation* recognizes both the inference and the curiosity that are involved in the promise of further information.

Although designers have written of Mystery, little empirical work has been done to ascertain circumstances in the natural setting that might enhance it. Gimblett, Itami, and Fitzgibbon (1985*) studied this question and reported that several factors were particularly effective in enhancing Mystery: screening, enclosure, physical accessibility (for example, a defined path), forest illumination, and a relative lack of "distance of view."

Perspective on Preference Matrix

The Preference Matrix evolved from extensive examination of the scenes that are most and least liked across numerous studies. It was through such observations that we were led to see that Complexity, the object of considerable study in nonenvironmental contexts, was not powerful in explaining preference for the natural environment. The most preferred scenes in many studies, however, reflect what we have called Mystery: There is partially hidden information, and something in the scene tempts one to explore further. Smooth ground textures with widely distributed trees that mark depth tend to be high in Legibility and are also among the preferred scenes. By contrast, disliked scenes often show relatively *little* Coherence or Complexity.

The pattern that such observations suggest is that preference calls for at least a modicum of the qualities that permit immediate processing. A *lack* of Coherence makes it difficult to understand what is before one; a *lack* of Complexity diminishes one's likelihood of becoming engaged in viewing. It is not necessarily the case, however, that preference is enhanced by having increasing amounts of these informational factors. For the two factors that rely on greater inference, however, there is the suggestion that "the more the merrier." With more Legibility, confidence is enhanced that the setting will continue to be understandable. More Mystery entices one to further exploration.

There is no attempt here to say that the Preference Matrix accounts for all aspects of environmental preference. Rather, it provides a framework that seems to be widely applicable. It points to the need for considering several factors as well as their combinations. What may seem to be a more parsimonious model, such as a continuum ranging from *unity* to *diversity* with the suggestion that most preferred areas fall somewhere along it, is far too simplistic. At the same time, a lengthy list of issues that pertain to scenic quality without any organizational guidelines is also not likely to be fruitful. Consideration of the needs to understand and to explore, however, provides substantial structure and permits analysis of preferences despite considerable human variation.

THE EFFECTIVENESS OF THE PREDICTORS

There are many ways to approach the "prediction" of preference – the question of whether one can reliably anticipate attributes that enhance people's preferences for particular settings. The two forms of analysis we have discussed so far – one based on preferences for perceptual categories and the other involving examination of the scenes with the highest and lowest preferences – have been helpful in suggesting some issues that underly preference. These analyses have led to insights about content concerns, openness and spatial definition, and the factors in the Preference Matrix as potential indicators of scenic quality or environmental preference.

The potential indicators, however, are derived from the interpretation of the results. In the discussion so far they have not been tested in their own right. We turn now to a third way to use the preference ratings in attempting to understand environmental preference. For this approach, the potential indicators are directly examined to see whether they do, in fact, help account for the preference ratings. To do this, each of the specific attributes considered to be an appropriate candidate for accounting for preference needs to be evaluated or rated for each scene in the study. These ratings are then related to the preference ratings.

In hindsight, one's research program would be quite different than it was. From one's current perspective, it would have been nice to have settled on a set of potential predictors or indicators, agreed on their definition, standardized the procedure for obtaining ratings, and then systematically studied a variety of environmental contexts. To the extent that following such an approach is an expression of *methodological preference,* it would seem to emphasize Coherence at the expense of the other aspects of the Preference Matrix! The actual direction this research program has taken can hardly be described as a straight and narrow path; it is an expression of the power of exploration in the search for understanding. It continues to be a framework that guides exploration, with no assurance of where the trail will end.

Early Research

Our first study in the environmental preference area (S. Kaplan et al., 1972*) was based on 56 scenes, carefully selected to reflect a continuum ranging from natural (with minimal evident human intrusion) to strongly human-influenced (with no evident natural elements). In addition to this major focus on content, the study was oriented to examining the role of the one factor that had been emphasized in the experimental aesthetics area, namely Complexity. Study participants were asked to rate each of the 56 scenes both in terms of this informational factor and to indicate how much they liked it. The correlation between these two ratings was .37, based on all the scenes.

Using the CIM approach, two perceptual categories

emerged that were strongly content-based. The scenes with clear human influence and virtually no natural elements formed the Urban category; those with natural elements (including many that showed human influence) formed the Nature category. (As mentioned in chapter 1, the scenes that showed residential arrangements did not form a clear content category.) Preferences for the two content domains were strikingly different, with the natural scenes far preferred. Clearly, content was an important element in preference.

Examination of the most and least preferred scenes provided further indication of the importance of content, as the most preferred scenes were of natural settings and the least preferred were urban. Examination of the most and least preferred scenes also made it evident that other factors should be taken into account besides Complexity. The very first version of the Preference Matrix was coming into being. Mystery and Coherence seemed to be important factors in explaining the patterns of preference.

We therefore used the same slides again and asked separate groups of participants to rate the scenes in terms of the two new informational factors that had seemed salient: Mystery and Coherence. (Legibility was not yet part of the Preference Matrix.) For purposes of replication, and because we were concerned about the potential biasing effect of rating scenes with respect to both an informational factor and preference, we also asked two other groups of participants to rate the scenes in terms of Complexity and preference. (Each person was asked to go over the scenes only with respect to a single rating.)

The results of this second effort (R. Kaplan, 1975*) were interesting in several respects. For preference and Complexity, the ratings that were repeated in the second study, the agreement between the original ratings and the new set was neither dismal nor encouraging. (In both cases, the correlation was .62.) Since participants in the original study had rated scenes in terms of both variables, these results suggested that ratings of preference may, in fact, have biased the Complexity reactions. This conclusion was reinforced by looking at some of the other findings. The correlation between the *independently derived* ratings of Complexity and preference was −.47, quite a contrast to the positive .37 when the ratings were done by the same individuals.

Thus we seem to have traded one perplexity for another.

It is now apparent that the positive relationship between Complexity and preference was an artifact; but why the negative relationship when the ratings were made independently? A clue as to what was going on here was provided by looking at the relationship between Complexity and preference separately for each of the two content domains that had emerged from the CIM approach. Preference for the Nature category was again far greater than for the Urban, as had been true in the original study (t = 4.22 and 6.76, $df = 32$, $p < .005$, respectively). Complexity, however, was far greater for the Urban than for the Nature scenes. (This was also true in the original study but not to the same extent: t = 7.08 and 3.13.) Using independent ratings, the correlation between Complexity and preference was not significant *within* either of the two content domains. Thus the overall negative correlation reflected *content,* the fact that the less liked urban scenes were more complex (Fig. 2.6).

What of the "new" predictors? Based on all 56 scenes, the correlation between preference ratings and Coherence was .34 and for Mystery it was .50. Ratings of Coherence showed no difference for the two content domains, but the

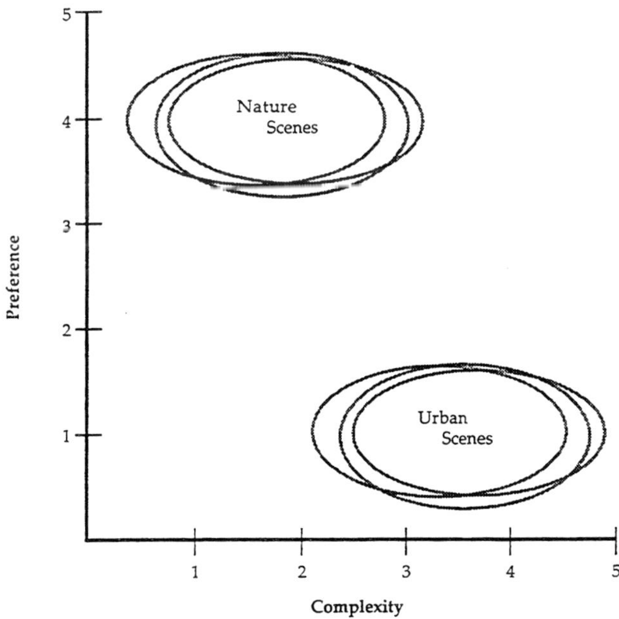

Figure 2.6 In a conceptual way, the diagram illustrates the relationship between Complexity and preference for two content domains.

Nature scenes were rated significantly higher on Mystery (t = 4.09). Among the 22 Nature scenes, preference was greater for scenes that had been rated as having more Mystery (and, to a lesser degree, this was also true for Coherence).

Although looking at the correlations is an intuitively satisfying approach to these data, a more appropriate procedure involves regression analysis. Such analyses look at the combined effectiveness of a series of predictors in explaining preference, summarized in a statistic called R^2. Part of the effectiveness of any particular indicator might be attributable to its joint effect with another indicator (to the extent that the predictors themselves may be interrelated). Regression analysis excludes this "double counting." Thus the effectiveness of each predictor is computed by excluding the potential effects of all the others included in the equation.

Using regression analysis, the results of this initial study (using independent ratings of the three informational factors) were very promising. R^2 was .49, indicating that the three factors together accounted for almost half the variance. Each of the three factors was significant, with Mystery the most powerful (partial r = .56), Coherence a somewhat weaker predictor (partial r = .33), and Complexity a negative factor (−.39). For the 22 Nature scenes taken by themselves, only Mystery and Coherence were significant predictors (R^2 = .42), whereas for the 12 Urban scenes by themselves none of the predictors was significant.

Later Studies

Several other studies have included ratings of informational factors, in addition to preference reactions. In most instances the ratings of the informational factors have been performed by a small group (four to eight individuals) who rate the set of scenes sequentially (often in different orders and over several sessions), each time considering a different factor. The intention is to have the panel members fully understand the meaning and nuances of the informational factors. In other cases, notably Herzog's studies (1984*, 1985*, 1987*, 1989*) and Herzog and Smith (1988), the predictors are rated by groups of students who are given no more than a simple explanation of the informational factor. Typically, in Herzog's studies any one participant is asked to rate the scenes in terms of a single factor or at most two of them.

The choice of predictors has also varied across studies. Mystery, Coherence, and Complexity have been included most frequently. Legibility was added to the Preference Matrix more recently and has thus received far less systematic attention. Other variables have also been included, notably Identifiability (Herzog, 1984*, 1985*, 1987*), Spaciousness (Anderson, 1978*; Gallagher, 1977*; Herzog, 1989*), Smoothness, Openness, and Ease of Locomotion (R. Kaplan et al., in press*), and variables drawn from Appleton's (1975) Prospect/Refuge model (Herzog, 1989*; Woodcock, 1982*). Though the predictors may have been called by the same name, the definitions that have been used have varied. Woodcock's use of Legibility, for example, comes much closer to Coherence as used in other studies.

The approaches to data analysis taken in the different studies have also varied widely. Herzog has generally examined the role of the predictors in the context of specific perceptual categories, where they can help interpret the basis for the category but are limited by the small number of scenes making up a category.[4] Other studies have generally reported either correlations or significant regression results, sometimes without indication of the direction of prediction.

The discussion here draws on a relatively small set of studies. Included are instances for which we have access to the original data and could, therefore, perform comparable analyses. The studies have in common that independent ratings were obtained on a series of informational factors. (Excluded are studies where the same participants were asked to rate the scenes in terms of their preference as well as any informational factors.) The studies include three doctoral dissertations (Anderson, 1978*; Gallagher, 1977*; and Woodcock, 1982*) and a master's thesis (Ellsworth, 1982*). In addition, five of Herzog's studies (Herzog, 1984*, 1985*, 1987*, 1989*, and with Kaplan & Kaplan, 1982*) and the R. Kaplan et al. (in press*) study are included in these analyses.

One would expect the informational predictors to be affected by the environmental context (or contexts) included in any particular study. The studies that are reviewed here varied widely in this respect. The studies by Gallagher,

Herzog (1989*) and Herzog et al. each included a mixture of built and natural settings. In Gallagher's case, the study area is a corporate headquarters (with the building visible in some scenes) and its surrounding lawn and landscaped areas. In the other two instances, the focus is on diverse urban places. In two studies, Ellsworth and Herzog (1985*), only waterscapes are included. The scenes in both the Anderson and Herzog (1984*) studies are of woods and fields; the scenes in the R. Kaplan et al. study include some of these settings, as well as agricultural land, in a more rural context. Unlike these other studies, which draw on a regional context, the scenes Woodcock used represent three different biomes and were taken in many places around the world.

It would also be reasonable to expect preferences to be affected by many issues other than those included in the Preference Matrix. In chapter 3, for example, we will consider the role of familiarity in preference. The studies under review here vary widely with respect to how much experience the viewers have had with the settings they are rating for preference. In Gallagher's study the ratings involve a local setting, with views that some of the employee participants had from their windows. Similarly, Anderson's participants were viewing scenes of the familiar local setting. In Woodcock's study, by contrast, it is quite unlikely that any of the participants had much experience with rain forests.

Given this assortment of studies and variations in all matters that may be pertinent to the question, can one say anything at all about the prediction of preference? Table 2.3 provides some interesting outcomes. It presents the results of regression analyses, showing the R^2 value as well as the partial correlation coefficient if it was significant (at $p < .05$). Results from the replication of the original study, discussed in the previous section, are included in the table as well. In all but the four Herzog studies (1984*, 1985*, 1987*, and 1989*), the regression analyses used only the informational factors in the Preference Matrix (Complexity, Coherence, Mystery, and Legibility), and these only if two or more were part of the study. In the case of the four Herzog studies the regression analyses also included other predictor variables, as indicated in the summaries of the studies in Appendix B, and this may partially account for the generally higher R^2 values indicated for these studies.

Table 2.3 *Comparison of results of regression analyses using informational factors in preference matrix.*

Study	No. scenes	R^2	F	p	Complex.	Coher	Mystery	Legib.
Woodcock	72	.19	8.28	.001	xx[a]	xx	.38	—[b]
R. Kaplan et al.	59	.19	3.11	.05	—	—	.31	—
Ellsworth	60	.24	4.35	.005	—	—	.31	—
Herzog, 1987	70	.50	10.64	.001	—	—	.39	xx
Gallagher	32	.42	6.75	.001	—	.51	—	xx
Herzog et al.	70	.13	3.25	.05	—	.26	.34	xx
Herzog, 1985	70	.46	9.11	.001	—	.39	.34	xx
Herzog, 1989	70	.87	43.70	.001	—	.60	.45	—
Anderson	48	.45	8.92	.001	—	.53	.49	−.43
Herzog, 1984	100	.53	17.42	.001	.33	.21	.33	xx
Replication	56	.49	16.76	.001	−.39	.33	.56	xx

[a]Not included in study.
[b]Not a significant factor.

Generalizing across these studies, the table suggests the following:

1. In each of the studies the combination of these informational predictors yielded significant results.
2. Complexity was a significant positive predictor in only a single study (and a negative predictor, as discussed earlier, in the initial study, where the Urban scenes were less preferred and rated higher in Complexity).
3. Legibility's role is hard to judge. In four of the five studies where it was included, Legibility did not play a significant role. In Anderson's study it was found to be a negative predictor.[5]
4. Coherence proved to be a significant predictor in the majority of the studies where it was included; in one case it was the only significant predictor in the regression analysis.
5. Finally, Mystery is the most consistent of the informational factors. (In the one study where it was not significant, the correlation between Mystery and preference [.45] was higher than the correlation between preference and either of the other two factors included in the study.)

Summary

The purpose of the Preference Matrix is to inform intuition. It is a framework, a structure for analysis. It suggests that the needs for understanding and for exploration are both important; one cannot replace the other. Similarly, the desire for both the immediate and the more inferential coexist. The results of the various regression analyses would suggest that on the understanding side, the immediately available information is particularly salient to preference. One wants to gain a quick grasp of the setting, and this is more readily achieved when it is coherent or well organized. As far as exploration is concerned, the emphasis seems to be on the promise of information, rather than on the information that is immediately available. Preference is greatly enhanced when the scene suggests that more could be learned from entering deeper into the setting.

It is important, however, to realize that these results cannot speak to the full intention of the framework. Regression analyses are useful for isolating the effectiveness of each factor in its "purest" form (i.e., having statistically eliminated potential joint effects with other factors). The Preference Matrix, however, is committed to the simultaneous necessity of several factors, not the optimization of one of them. It would be most unfortunate to conclude from the regression analyses that the maximization of Mystery and the neglect of the other aspects are sufficient in designing a highly preferred setting.

We also want to emphasize that the framework proposed here is a conceptual tool that is guided by new information and insights. It is a model that continues to undergo change. Legibility in particular requires further development. Its basic intent seems solid: Environments that afford a good setting for functioning in a competent, comfortable, and safe fashion should be more preferred. How this conceptualization manifests itself in diverse settings remains a challenge for further research.

PREFERENCE AND THE NATURAL ENVIRONMENT

People, like trees and snowflakes, differ from each other. They see the world through (conceptually) different eyes and bring diverse backgrounds to any new experience. People vary as well in what they like and dislike. Given this wondrous diversity, it is just as wondrous to find some strong and pervasive consistencies in the way people interpret the environment and in their preferences.

The strong preference for the natural world is a basic theme of this book. We see evidence of it in this chapter in the higher preferences for scenes that reflect nature as opposed to more human-influenced elements. The preference for and importance of trees are evident in the results discussed in this chapter and will receive further corroboration in chapter 5. The sensitivity to natural elements and the preference for them have now been substantiated in numerous studies (see Schroeder, 1988, for an excellent review).

That is not to say, however, that all that is natural is preferred. The challenge, then, is to determine what differentiates the natural areas that are more favored from those that are not. As we have seen, examination of most and least preferred scenes in any particular study provides some insights into this. Analysis of the relative preferences for empirically determined perceptual categories also sheds light on this issue. Finally, the studies that systematically assess attributes that are potential candidates for "preference predictors" offer yet another way to approach this concern.

From these very different forms of analysis come some very consistent conclusions. Preference seems to be intimately related to effective functioning. This relationship is generally not a result of conscious calculation. In fact, people are often willing to indicate their preferences based on the briefest glimpse, permitting no opportunity for careful reflection. Yet, ratings based on the briefest durations seem to be similar to those made in a less hasty fashion.[6] Of course, such preference judgments are made by all of us very frequently, and not only in a research context. That one should favor a path - literally and figuratively - that facilitates functioning is hardly surprising.

The next step in the quest, then, is to identify some of the properties that characterize effective functioning. In an environment that fosters effective functioning, one might expect the individual to experience a sense of both safety and competence. One might also expect that individual to feel reasonably comfortable about the situation. These three aspects of an individual's feelings about an environment are admittedly far from pure; they undoubtedly overlap considerably. Nonetheless, they provide a useful, if rough-and-ready, indicator of the impacts of different sorts of environments. If the environment can make these more likely, one would expect it to be preferred. It is thus interesting to examine the qualities of preferred natural settings in terms of their relationship to these aspects of functioning.

We have seen that across very different environmental settings, scenes that are extremely open and lack any elements that offer help in differentiating the depicted space are relatively lower in preference. Although such scenes might contribute to a sense of safety because it is easy to see in all directions, they offer little protection should one need it. Competence is undermined by the lack of landmarks and other features that help one maintain orientation. Finally, the prospects for continued competence are likely to be reduced, given the limited opportunity for exploration.

Preference is also consistently low in settings that are blocked. A dense tangle of understory vegetation dominating the foreground of a scene, for example, makes it unlikely that one could move readily through the area, thus reducing one's competence. The sense of lurking danger, given the obstructed view, would also make the setting seem less safe and less comfortable.

Relatively uniform and short ground textures enhance the ability to locomote. Being able to move readily through an area gives one a greater sense of security and competence. Even if one cannot move easily through a setting, the opportunity to see through the trees can serve some of these needs. If there is visual access, the assessment of one's ability to function is also likely to be enhanced. It is far easier to avoid being surprised in an open, transparent forest than in a dark and dense one.

Coherent, orderly settings are likely to increase one's sense of competence because they are readily graspable. One can more readily determine where to focus attention and to sense what might lie ahead. Similarly, settings that are legible provide cues that help in maintaining orientation, thus making safety, competence, and comfort more likely. The lack of Complexity suggests a setting that has too little to structure it, as is true of the wide-open areas. In a setting that has Mystery, one can learn more in a cognitively comfortable and safe fashion.

From these brief descriptions, it becomes apparent why the informational factors in the Preference Matrix help explain characteristics of preferred environments. The spatial definition or structure of an area, the textures that help one decide about the ease of locomotion or visual access, and the invitation to enter the scene to learn what cannot be determined from one's present vantage point are all powerful yet subtle qualities of the environment. Without realizing it, humans interpret the environment in terms of their needs and prefer settings in which they are likely to function more effectively.

NOTES

1. *Habitat selection* is the technical name given to the inclination to prefer environments that make successful adaptation more likely. Among vertebrates, habitat selection is a widespread tendency (Woodcock, 1982*). What this means is that animals tend to show a preference for the kind of environments in which their species prospers. In some instances this occurs even if the animals have been raised in the laboratory and have had no direct prior experience with the environment in question (Wecker, 1964).
2. The original version of the Preference Matrix was included in

S. Kaplan and Wendt (1972) and subsequently modified (S. Kaplan, 1975, 1979b). In those papers, Making Sense and Involvement were the labels for the informational domain. The use of Understanding and Exploration for these concepts began with S. Kaplan (1987a).

3. A vivid demonstration of the relationship between a failure to understand and an emotional reaction can be found in the classical study of Bruner and Postman (1949). In this fairly typical exploration of the human perceptual process, an image was flashed on a screen and the task was to identify it. If the image was not identified at the initial duration, say 1/100 of a second, then it was projected for a slightly longer duration, and so on until it was correctly identified. In this particular study the images were playing cards; to make the task a bit more challenging, they were printed in the opposite color from the usual pattern. In other words, spades were red and hearts were black. The performance of one particular individual on this task was especially striking. The duration had to be increased steadily because of the participant's inability to identify the pattern. Finally, the image appeared on the screen for 0.3 sec, quite a long time for a study of this kind. The response to this extended duration was as follows: "I can't make the suit out, whatever it is. It didn't even look like a card that time; I don't know what color it is now or whether it's a spade or a heart. I'm not even sure now what a spade looks like! My God!" One must admit that this is a rather trivial situation. Nothing terrible is going to happen if the pattern is not correctly identified. Yet the concern to comprehend, to make sense of the world around one, is urgent no matter what the situation.
4. Herzog has provided us information to include in the analyses here that, in some cases, does not appear in the published work. In other instances too, the material included in this book, though referring to published works, includes analyses that were carried out for the present purposes.
5. There are theoretical grounds for considering Identifiability as a component of Legibility. Herzog has explored this predictor in several studies. In the 1985* study, Identifiability was not a significant predictor, whereas in both the 1984* and 1987* studies it was. In the Herzog, Kaplan, and Kaplan (1982*) study, an interaction between Identifiability and the content of the scenes led to a significant negative prediction. The scenes with strongest Identifiability, the Alley/Factory category, were also the least liked.
6. Comparison of preference ratings when the scenes are presented for 15 sec as opposed to very brief durations (e.g., 10 msec) have not been completely consistent. In some cases, preferences have been more extreme (disliked categories receiving even lower ratings and like categories receiving higher ratings) when the scenes were seen only briefly (Her-

zog, 1987*; R. Kaplan, 1975*). In other cases,the opposite pattern was found (Herzog, 1985*). In still other cases, the means were either higher (Herzog, 1989*; Herzog, Kaplan, & Kaplan, 1982*) or lower (Herzog, 1984*) at the short durations. Whether these preferences are statistically significant or not, a more important point is that even at the very shortest durations individuals have no difficulty in making these judgments, and they are substantially no different in their preference than when the scene is available for a prolonged time. Given that judgments of the environment are often made based on no more than a glimpse of the passing scene, this pattern of findings is not really surprising.

CHAPTER 3

VARIATIONS: GROUP DIFFERENCES

SEVERAL decades ago Florence Kluckhohn (1953) suggested certain "basic human problems" that must be addressed by all cultures. Her thesis was that each culture develops a dominant solution to each of these concerns but that the solutions also show identifiable variations for specific groups within the culture. One of the themes Kluckhohn included in her short list of basic human problems was "man's relation to nature." All cultures must address this question, but within any culture there will be dominant and variant values attached to it.

Inclusion of this theme as a basic human problem is in itself noteworthy. Kluckhohn's interest in this issue was mostly from the perspective of the balance or domination of humans and nature. From the material presented in the previous two chapters it would seem that, taken more generally, the role of nature is, indeed, a basic human theme. Chapter 1, on categorization, provided some interesting support for the relevance of the balance between the natural and the built in the way the environment is experienced.

The material in chapter 2, on preference, documented the consistent findings of strong preference for natural settings. In fact, the literature in environmental preference is rich with examples of consistency. Studies using diverse environmental contexts, varying ways of presenting information about the environments, and groups differing in various background characteristics have repeatedly reported high levels of agreement in preference ratings. In certain respects these similarities may be even more important than the differences that are the focus of this chapter. The fact that such consistencies have been found across many studies supports the notion of dominant values with respect to the role of nature.

Our attention in this chapter turns to the variations on the theme of preference. That there should be variation in preference must come as no surprise. In fact, preferences are often thought to be completely idiosyncratic – "there is no accounting for taste." Complete idiosyncrasy, however, has hardly been evident in the research on environmental preference. Rather, the pattern is one of strong continuities and similarities. At the same time, however, there are some important differences.

The major factor accounting for differences is familiarity. Familiarity is the product of experience, and experience comes in many forms. One gains such familiarity from many circumstances, such as where one lives, where one has visited, what one has studied, and the cultural norms of one's group. Some of the ways in which these circumstances affect preferences are explored in this chapter. In examining these findings themes from the previous chapters appear again. Thus content and spatial configuration are still dominant issues; understanding and exploration also still provide a useful framework. The relative importance or weighting of these different aspects of preference, however, changes.

Though familiarity is a major factor in understanding these variations, familiarity is by no means a simple predictor of preference. The notion that familiarity breeds contempt has truth value. So does the notion that one prefers the things one knows. It is not the case, therefore, that simply knowing that a particular group has greater familiarity with some environmental settings will be sufficient basis for understanding differences in preferences. On the other hand, recognition of differences among groups and differences between experts and nonexperts is vital if one is to design and maintain natural settings for diverse users.

The chapter is divided into three main sections reflecting different sources of variation in preferences. The first explores familiarity or experience that is related to the geographic circumstances of residence and to the effect of direct exposure to an environment. The second section looks at cultural and ethnic variation, including the question of age and other bases for belonging to a "subculture." The third section examines the effects of formal knowledge and expertise on environmental preference and perception.

There are two disadvantages to this organization of the material. The first is that the three categories overlap; the second is that a particular study might be pertinent to more than one of the three categories. In most cases, studies that are discussed repeatedly are included in Appendix B (and are designated by * in the text) so that a more coherent understanding of the study can be obtained in this fashion. Despite these disadvantages, however, we hope this organization might make an intricate and potentially confusing body of material more meaningful.

FAMILIARITY WITH THE LOCAL ENVIRONMENT

It would be reasonable to expect that the experience of rural living would affect preferences for natural settings differently than urban or suburban residential patterns. The view of nature is distinctly different in each of these contexts, and the people seeking one or another of these settings may have different nature preferences to begin with. Despite these considerations, however, studies that have explored this background issue have generally found no relationship that is clearly attributable to the urban/rural experience of the participants.[1]

Such findings may be surprising, but there are many reasonable explanations. Where one has lived for periods of time may provide a poor indication of familiarity with other kinds of settings. One also may not live in the kind of setting one would choose if one had such luxury. Furthermore, the very categorization of *rural, urban,* and *suburban* may not usefully parallel the way people experience different kinds of natural settings. Such groupings of the environment may be appropriate from a land-use perspective but may be far less useful as typologies of natural settings.

Where One Lives

By contrast, some consistent findings suggest that people relate differently to settings with which they have direct experience. The environment near one's home, for example, holds some special significance that is reflected in preference judgments. This is an area of great familiarity; it also is related to a vast array of issues that matter to the individual.

The Swift Run Drain Study (R. Kaplan, 1977a*) provides

a direct analysis of the relationship between preference and place of residence. The 5-mile stretch of this storm water drain includes a variety of residential areas. The character of the waterway itself ranges from a relatively unmanaged, "wild" appearance to regions where mowed lawn reaches to the edge of the water in relatively kempt fashion to areas where the water is fenced off and the adjacent land neglected (Fig. 3.1). Along one of its six distinct regions the drain is buried underground, the only evidence of its existence a row of mature willow trees in a flood plain that serves as backyards for rented dwellings. The photo-questionnaire used in this study sampled from each of these portions of the drain and also included scenes taken elsewhere to represent other potential problems and treatments of a storm drain system.

For each of the 32 photographs, participants indicated both their preference and how similar the scene is to their

Figure 3.1 Scenes showing the variety of regions along the Swift Run Drain. These were all included in the photo-questionnaire (R. Kaplan, 1977a*).

view of the waterway. Preferences for these scenes varied widely across the different segments of the drain. Preferences were not, however, consistently greater for the more similar scenes. Where the scenes were low in preference, they were equally low or even lower in areas where similarity was high (Fig. 3.2). In other words, an undesirable situation is not made less distasteful by its familiarity.

In other cases the average preference rating was clearly highest in the area where the view was most familiar. Four scenes showed extreme variation in preference, a range of 2.3 or more (using a 5-point rating scale) across the six regions. Three of these (Fig. 3.3) depicted grassy areas with willow trees and no visible drain. These views were highly preferred by the residents living along the stretch of Swift

Figure 3.2 These scenes were rated low in preference whether or not participants were familiar with them. The scenes in the top row were, in fact, characteristic of the drain where some participants lived. The lower left scene, by contrast, was taken in another city and was thus unfamiliar to everyone. (Equally unfamiliar, and taken in another city, was the scene that was most preferred by the study participants.)

Figure 3.3 Scenes where the drain is underground. These were far more preferred by the adjacent households than by other study participants.

Run where it is buried. Their ratings for these scenes were more than a full scale point greater than the ratings at any other region; across these other regions the variation in preference was also quite large. The fourth scene (Fig. 3.1, upper left) had a mean rating of 4.6 by those living nearest it. Residents of other regions rated it between 2.0 and 3.1. We shall return to this scene later in the chapter.

The relationship between preference and familiarity with the nearby-natural environment was also examined in the photo-questionnaire study of multiple-family housing (R. Kaplan, 1985a*). Here residents indicated the availability of a setting such as the one pictured as well as their preference for it. Two of the perceptual categories derived from the preference ratings consisted of natural settings (discussed in chapter 1; see Fig. 1.3). Such natural areas were most preferred and low in availability. The least preferred categories showed buildings dominating the landscape (Fig. 1.2, top). These were midrange in availability. There were two additional categories, both midrange in preference. One of these was rated as within walking distance by most participants (top row, Fig. 3.4), whereas the other was as unavailable as the nature scenes (bottom row, Fig. 3.4). Figure 3.5 summarizes these findings. Certainly, this pattern of findings suggests that preferences are not a simple outcome of what one can experience most readily.

Participants in this study also indicated how readily they could see certain features from their residence and how satisfying they found these views. Here again we find a complex pattern (Fig. 3.6). Views that are highly familiar (parking areas, large mowed expanses, trees, and landscaping) vary greatly in preference. At the same time, views that are highly preferred (woods, gardens, large trees, and landscaping) may not be very available. Absence

Figure 3.4 Top row shows scenes from Open, Residential category; bottom row scenes were in Landscaped category (R. Kaplan, 1985a*).

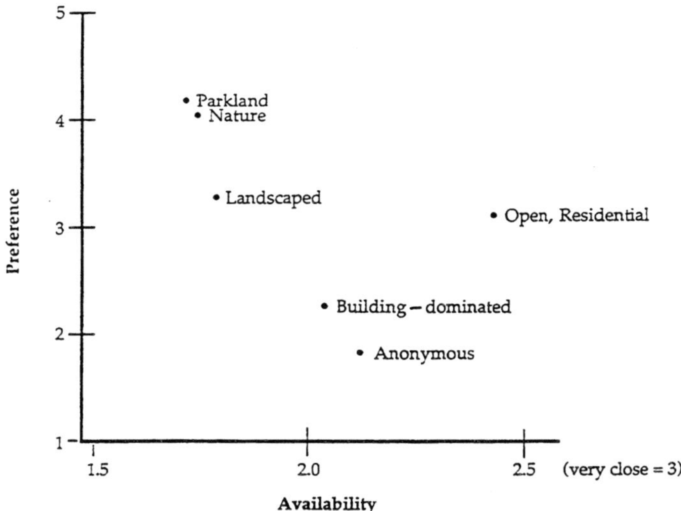

Figure 3.5 Relationship between Availability and Preference for photograph-based groupings.

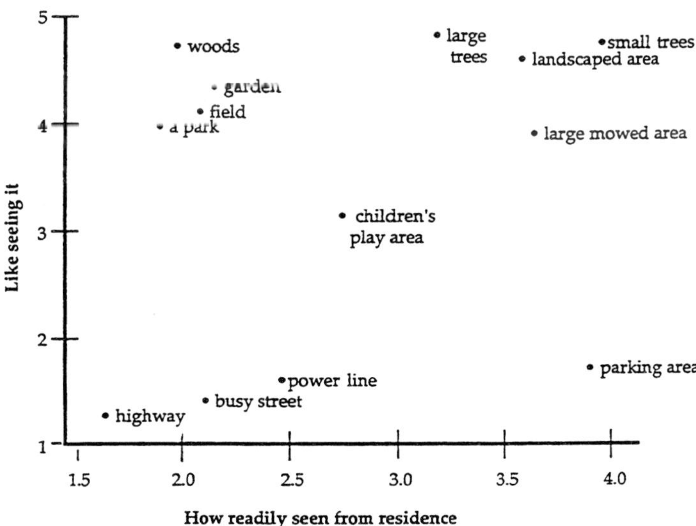

Figure 3.6 Relationship between dominance of view and satisfaction with view, based on verbal items in questionnaire.

may indeed make the heart grow fonder, but presence need not mitigate appreciation.

Two other studies shed some light on the importance of place of residence in relation to preference. In both cases a comparison is afforded between place of residence and place of work. In these examples, familiarity may be relatively comparable, but other factors would seem to affect the judgments.

Gallagher's (1977*) study included both employees of a commercial facility and residents living near it. The 15 acres of grounds surrounding the facility, CUNA headquarters in Madison, Wisconsin, had been redesigned, starting in 1972, to include landscaping that requires less direct management - a prairie restoration using natural landscaping principles. The photo-questionnaire consisted exclusively of views taken on the grounds of the facility, including areas that were kept as lawns, prairie areas with their tall grasses (Fig. 2.4), and regions to which some of the plant material originally included in the landscaping had been moved. These latter examples had relatively more trees (see Fig. 1.7).

Gallagher found the employees and neighbors differed significantly in their preference for the prairie scenes, with the neighbors showing greater appreciation. Among the neighbors, those living in the adjacent apartments were much more favorable toward the prairie restoration than were the nearby homeowners; the latter were significantly more favorable toward the views of lawns. In addition to the preference differences, Gallagher also reported striking differences in the correlation between preference predictors and preference ratings. Coherence was a negative predictor of preference for the apartment dwellers (−.44) whereas it was positive for the homeowners and employees (.52 and .42). Similarly, the role of what Gallagher called "naturalness" (the unmanaged appearance) was opposite for these groups (.49, −.56, and −.37). Mystery and the role of trees were equivalent in all cases.

Several factors may help explain the differences between the two neighbor groups. Gallagher's sample of apartment dwellers was very small, and he is cautious in interpreting the results. However, these individuals were very favorable to the natural, unmanaged area. The homeowners, by contrast, remarked on the "messy" appearance. Their own residences tended to be landscaped traditionally with large lawns and a specimen tree. It is not surprising that homeowners and renters would have different concerns for the appearance of the area immediately surrounding their residences. The fact that the apartments were relatively expensive may also play a role; the concern for order in the nearby-natural environment seems to be somewhat less salient at higher socioeconomic levels. Furthermore, the availability of nearby open space was a major reason for choosing these "luxury" apartments for many of their occupants.

The second study that permits comparison of place of residence and place of work involved public input in early design phases for a downtown vest-pocket park (S. Kaplan & Kaplan, 1989). Because the input was gained prior to construction, physical models were used for simulation, and these were kept relatively low in detail (Fig. 3.7). Participants were asked to rate their preference for each of 24 photographs of these simulated park environments. They

Figure 3.7 Photographs of simulations used in a citizen survey of a proposed small urban park (S. Kaplan & Kaplan, 1989). The models and these photographs are the work of Terry Brown and Charles Cares, the designers of the project.

were also asked whether they lived and/or worked in an area near the proposed park.

The differences in preference between the people who work downtown and those who live there were substantial and consistent. Individuals who work in the area were much more positive in their ratings. For them the park, once completed, would offer a lunchtime respite and a pleasant oasis while running errands. For the nearby residents, however, the proposed park was viewed with some trepidation. Would undersirable individuals and "muggers" hide behind bushes and structures? The scenes were interpreted with such concerns in mind (and, in many cases, expressed). The park designers attempted to deal with these concerns in their plan for the park. They were apparently successful in doing this; some years later when the park was evaluated after it had been in use, the local residents and nearby workers were equivalent in their praise and did not differ in their judgments of the safety of the park (R. Kaplan, 1980).

Locals and Tourists

A similar contrast to the work–residence distinction can be made between local groups and occasional visitors. Here again there are likely to be differences in degree of familiarity, but other factors must also affect one's view of home range.

A study of scenic routes in Michigan's Upper Peninsula (R. Kaplan, 1977a*) used a photo album with each page showing four different views one might come upon while driving through the region. Participants were asked to evaluate the foursome in terms of preference. The sample included both year-round residents of the area and individuals who happened to be visiting the region. Over all the scenes, the locals showed significantly lower preference than did the tourists. The locals, however, were more differentiated in their preferences (Fig. 3.8). They greatly preferred scenes of open forests to dense forests, whereas the visitors much preferred forests to flat farmland, without distinguishing the forest density in preferences.

Miller's (1984*) sample consisted of ferry passengers who were asked to indicate their preferences for views of the British Columbia coastline. One of the groupings that

Figure 3.8 Locals and visitors to the area differed in their preferences for scenes such as these from Michigan's Upper Peninsula.

emerged from the Category-Identifying Methodology was Clearcuts and Natural Bald Spots. These scenes were significantly more preferred by passengers residing outside British Columbia (visitors) than by passengers whose residence is more local. Out-of-province participants were less likely to recognize these scenes as showing timber harvesting. The local residents, by contrast, showed greater appreciation of the Low Flat Shorelines grouping. Here again familiarity may play a role, with the tourists' expectation for shoreline topography negatively impacting their preferences when shown scenes lacking in relief.

These studies are interesting both for the conclusions one can draw and for the absence of regularities where one might have expected them. It is clear that the local population is likely to have strong feelings about the local landscape. It is also clear that there are important areas of differences between locals and visitors. Attracting tourists may come at the cost of the affections of nearby residents. One conclusion that one *cannot* make is that the very familiar environment is necessarily more preferred. The characteristic flat farmland scenes of the Upper Peninsula were more preferred by tourists than by locals, whereas in Miller's study the flat shoreline scenes were more preferred by the provincial residents. Given the predictably strong concern of the locals, the potential for conflict between local and visitor preferences, and the difficulty of predicting the specific pattern of preferences for any particular locale, obtaining preference data at the local level would appear to be essential.[2]

Nature Hikers

A very direct way of exploring the issue of familiarity or experience is to study preferences both before and after a visit to a particular site. If individuals have never before been to the site, the question of familiarity is relatively well controlled and changes in preference can more easily be attributed to the experience.

As part of the Outdoor Challenge program (discussed more fully in chapter 4), participants were asked to rate 24 scenes in a photo-questionnaire (R. Kaplan, 1984b*). Two sets of ratings were printed beneath the photographs: "How familiar are you with scenes like the one shown in the picture?" And "How much do you like the scene?" All but three of these scenes reflected the various forested environments encountered in the course of the program. Included were relatively dense forests, open woods, swamps, and rocky areas. None of the scenes would be considered particularly scenic. The three "control" scenes, those not taken in the area of the wilderness trip, depicted common roadside scenery, showing relatively open country with some trees and coarse-textured vegetation in the foreground.

These ratings were completed on two occasions: immediately prior to departure on the 9-day arduous wilderness trip and at its conclusion. The results are based on 49 participants who went on the trip in eight different groups in the course of two summers. Of these, 22 were adult participants (17 women and 5 men, ages 19–48) and 12 boys and 15 girls who were high school students.

Familiarity ratings were significantly greater at trip's

end than before departure for all but three scenes. These were the three scenes that were not taken in the area of the wilderness trip. Thus changes in the familiarity ratings provide a strong indication that particpants were aware of the kinds of habitats and ecosystems through which they traveled.

With the exception of four scenes, preference ratings did not change significantly from the initial to the final ratings. The only scenes that showed significant preference changes were the four comprising the Swamp grouping (based on CIM). For these, preference declined (from 3.6 to 3.1, on a 5-point scale). Here much experience with traversing swampy areas seemed to have bred a certain amount of distaste for such habitats.

Both Hammitt (1978*, 1987) and his student Keyes (1984) have studied the relationship between familiarity and preference with respect to a short hiking experience. The brevity of the hike presents the problem of asking individuals to rate the same scenes in a matter of a few hours. The various studies in this series have approached this issue in a variety of ways, permitting comparison of ratings before and after the hike by different individuals, comparison of preference and familiarity ratings by same or different individuals, and comparison of ratings by same individuals using partially overlapping sets of photographs. Each of the studies has been noteworthy for the large sample sizes.

Several findings have been consistent across these studies. (1) Even the single hike along a nature trail (in Cranberry Glades, West Virginia, or the Trillium Gap Trail in the Great Smokies) leads to extremely high familiarity ratings. Individuals are clearly aware of the scenes along the trail and recognize them in the photographs. (2) Both preference and familiarity ratings by different groups show remarkable consistency. (3) The relationship between preference and familiarity is generally positive, but there are also very memorable scenes that are not preferred. A scene of an uprooted tree, for example, receives low ratings on preference but high ratings on familiarity.

Summary

Direct experience and knowledge of a place can clearly affect preference. It is less clear, or predictable, what the effect will be. Visitors very rapidly become familiar with an area and can recognize scenes of it. Nonetheless, the experiences of longtime residents of an area reflect both greater differentiation of the landscape features and different appreciation of the common aspects of the landscape when compared to visitors to an area. Furthermore, residents are likely to be attached to the nearby-natural environment in ways that are quite different from more transient users, such as people who work in the area.

Based on these aspects of familiarity or experience, however, it is difficult to say how preference is affected. If anything, the pattern of these results simply suggests that before changing the nearby-natural setting one might better determine what the locals like and dislike about it. It would be difficult for a planner or designer to foretell the consequences of change given the results reviewed thus far.

CULTURE, SUBCULTURE, AND ETHNICITY

Another way to think about "where one lives" is at a cultural, as opposed to individual, level. People who share a system of thought or perhaps a language (or dialect) and pass these along from generation to generation would presumably experience the environment similarly and might have some common preferences. In addition, people who share an environment would presumably have greater familiarity with its vegetation patterns and seasonal effects, which would be expected to influence their perceptions and preferences.

For purposes of our discussion it is not important to draw clear lines between culture, subculture, and ethnicity. It is useful, however, to explore some of the shared preferences and perceptions of such groups. Studies have touched on these sources of variation, but the results must be viewed with caution. These studies have generally not been carried out in a way that would give one great confidence in generalizing widely. On the other hand, it is likely that insights about subcultural influences will be forthcoming only through the accumulation of such small-scale studies. Consistency across these efforts is well worth heeding.

Cross-cultural Comparison

The literature on cross-cultural comparison in preference for natural environments shows relatively high agreement when cultures are similar (Zube, 1984). Zube and his students have reported correlations between .76 and .89 among two Australian samples and American landscape architecture students, using photographs of Australian landscapes (Zube & Mills, 1976), and similarly high correlations in comparisons among Yugoslavian students, Italian-Americans, and various American groups, using scenes of northeastern American landscapes (Zube & Pitt, 1981). Others have reported consistent preferences when comparing Americans with Scots (Shafer & Tooby, 1973) or Swedes (Ulrich, 1983). These studies, however, have focused largely on correlational results.

In some of our recent work in this area we have found that high correlations between groups tell only part of the story. Though the groups may agree in their *relative* likes and dislikes of certain scenes (reflected in high correlations), they may nonetheless differ substantially in how much they actually like the scenes (mean rating). Furthermore, despite high correlations, it is possible that the groups differ in how they experience the environment, as reflected in perceptual categorizations (CIM-based results). Such comparisons across different analytic methods are not available for the studies reported in the previous paragraph. Let us turn, then, to three studies that permit such exploration. Across these, the cultural influences that are examined entail both the landscape and the participants. In Yang's (1988*) study, the focus is on oriental and western influences, whereas in the R. Kaplan and Herbert (1987*, 1988*) studies the comparison is between Western Australia and the north-central United States.

Yang's study is one of very few in the preference literature that examines non-western landscapes; it is also unusual for its samples. All 40 scenes in his photo-questionnaire were taken in Korea, but they were selected to reflect Korean, Japanese, and western-style landscapes. (Korean-designed landscapes are characterized by fences, walls, and pillars, all constructed of quarried stone blocks. Rectilinear rather than curvilinear forms tend to dominate the highly structured settings.) For each style, distinct landscape characteristics (water, vegetation, rock, and layout

Figure 3.9 Scenes in top row were among the most preferred in Yang's (1988*) study for both Korean and western samples. Both are waterscapes: *left,* Japanese-style, *right,* Korean-style. Scene at bottom left is designated western-style and was among the most preferred for the Korean sample; scene at bottom right is an example of Korean style and was among the six top scenes for the western sample. Neither of these latter scenes is a waterscape, and each shows preference for a "nonnative" style.

of space) were included. Both Koreans (560 citizens and students) and western tourists (N = 110) served as participants.

Although the tourist sample's preference ratings were significantly higher (mean 3.3 as opposed to 2.8 for the Koreans), the relative preferences showed strong similarities. Of the six most preferred scenes for each group, four were the same (Fig. 3.9), and water was a dominant theme among the most liked scenes. The least preferred scenes also showed overlap (three scenes common to the set of six for each sample), and the use of quarried rock was the dominant theme among the less preferred landscapes (Fig. 3.10). The samples were also similar both in their preference for Japanese landscape style and in their relatively lower appreciation for the style representing their own culture.

Figure 3.10 *Top row:* scenes that were among the six least preferred for both samples in Yang's (1988*) study. *Bottom row:* scene at left was among least preferred by Koreans though not by western sample; scene at right was low in preference for the western sample but not for Koreans. All four scenes are examples of Korean style, and all include the quarried rock element. Scene at lower right provides a relatively unusual example of a water scene that received a low preference rating.

CIM procedures generated four categories for each sample, and these showed both similarities and differences. The most preferred category in each case reflected the preference for water and for Japanese style, with six common scenes. In the case of the Korean-based category, additional scenes suggested that the common theme for that grouping was Water Surrounded by Vegetation. The scenes included only in the western-based grouping suggested naming the category Japanese-style Landscapes. The least preferred categories showed a similar pattern, with nine common scenes. The additional scenes for the Korean-based grouping more strongly reflected Rock with Sparse Vegetation, whereas the scenes unique to the western-based category were more uniform with respect to the style, and hence the grouping was named Korean-style Landscapes.

The other two categories for each sample were based to a large extent on the differences between western formal and western informal style, although not all the "informal" scenes for one sample were, in fact, designated "western." For both groups the informal category was significantly preferred.

The strong cross-cultural similarities in Yang's study are surprising in some respects. One might certainly have expected stronger preferences on the part of the Koreans for their own, dominant, style. On the other hand, the Korean sample's perceptual categories do not distinguish as clearly between the two oriental styles, as did the western sample. The designation of "western" style might seem misleading to a western researcher, who would be likely to see distinct differences among these. The empirically generated categories made it evident that the participants, regardless of cultural orientation, also did not perceive this as a distinct style.

The other two studies (R. Kaplan & Herbert, 1987*, 1988*) involved university students in the United States and in Western Australia.[3] Though all the scenes were of nondramatic natural landscapes, in one instance the scenes were taken in Michigan and in the other Western Australian landscapes were used. First-year students at the University of Michigan and at the University of Western Australia are culturally distinct, though the cultures are relatively similar. On the other hand, the characteristic vegetation in the two instances is distinctly different.

The 55 slides taken in the rural western parts of Oakland

County, Michigan, included scenery typical of midwestern and eastern portions of the United States, consisting of abandoned fields, forests, and rural housing. There is little topographic variation in the region (Herbert, 1981*). The 60 slides of Western Australia sampled five landscape types found in the northern Jarrah forest: escarpment, uplands, valleys, plantations, and eastern woodlands. The Australian and American scenes contrast in two important respects: the relative aridity of many of the Australian settings and the characteristic gray-green foliage of the dominant eucalypt species of the Australian landscape.

The results support previous findings when viewed in terms of correlations between the two samples. Correlation of .84 was obtained in each study. The mean ratings in the two studies show that each group favors its own landscape slightly (but significantly), with means of 3.3 and 3.1 across the respective slide sets. For the Australian landscapes there were four instances where the American students significantly preferred a scene. Two of these were of vistas (one with a river dominating the scene) that could have been taken in many parts of the United States. The other two showed a palm tree in one instance and bare trees in the other. Since bare trees are more likely to reflect disease than season in the Western Australian context, this difference is understandable. The palm tree was presumably more familiar to the American sample than the eucalypts visible in most of the scenes. These results suggest relatively high agreement between the two groups and that familiarity enhances preference to some degree.

The comparison between the two groups turns out to be more interesting, however. If we use the Category-Identifying Methodology (CIM) separately for each cultural group in each study, some differences in both perception and preference become evident.

For the American scenes (R. Kaplan & Herbert, 1988*), the groupings were very similar and the preferences equivalent. Table 3.1 shows that the four categories generated by the CIM procedure can be described using the same label even though the scenes comprising the categories are somewhat different using the American-based and the Australian-based ratings.

The major differences between the two analyses involved somewhat different interpretations of the Vegetation and the Pastoral groupings. The Australian-based Vegetation grouping included 12 of the 17 scenes in the

Table 3.1 *Preference means for the American scenes based on American (UM) and Australian (UWA) data sets (R. Kaplan & Herbert, 1988*)*

	UM-based analysis			UWA-based analysis		
	No. scenes	UM	UWA	No. scenes	UM	UWA
Manicured/Picturebook	4	3.93 (1)	3.83 (1)	3	3.90 (1)	3.80 (1)
Vegetation	17	3.52 (2)	3.42 (2)	14	3.48 (2)	3.44 (2)
Pastoral/Open Rural	15	3.07 (3)	2.64 (4)	21	3.12 (3)	2.70 (4)
Residential	10	2.81 (4)	2.79 (3)	9	2.91 (4)	2.91 (3)

Note: Numbers in parentheses represent rank order of preferences within a column.

American-based version. Dropped were five scenes that were distinctly less green than the rest, and two grassy scenes were added. Thus, though the two Vegetation groupings are very similar, the one based on the Australian sample has a stronger sense of lush, green landscapes and more forested settings (Fig. 3.11). The preferences were essentially identical.

The scenes comprising the Pastoral category were quite different when based on the ratings by the Australian as opposed to the American students. Thirteen of 15 of the scenes in the American-based category are included, but eight are added in the Australian version. With these additions, the grouping is more suggestive of open, rural land than of a pastoral setting (Fig. 3.12). Whether using the Pastoral or the Open Rural version of this grouping, the mean rating is distinctly lower for the Australians (2.6)

Figure 3.11 Differences between American-based and Australian-based Vegetation category (R. Kaplan & Herbert, 1988*). Scenes in the top row were common to both the American-based and the Australian-based CIM results. Those in the bottom row were unique to each of the analyses.

Figure 3.12 Differences between American-based and Australian-based Pastoral category (R. Kaplan & Herbert 1988*). Top row scenes were included in both instances. The bottom row, however, represents the different perception of the Pastoral grouping to reflect a more Open Rural landscape pattern on the part of the Australian sample.

than for the Americans (3.1). Although these scenes are far preferred to the Residential category by the American students, for the Australians they are the least preferred of the four clusters.

The conclusion that Americans prefer their more familiar landscape is quite misleading given these findings. It seems that, although the Australians are not directly familiar with any of the scenes, the genre of the open rural setting is one they recognize readily. This is probably the most common nearby nonurban landscape in the area near Perth, Western Australia. In this case, a relative sense of familiarity for settings that are not highly preferred seems to breed contempt. For the remaining scenes, the two groups are equivalent in preference. Whether this means that the relative novelty of lusher landscapes enhances the

Australian students' preferences cannot be determined from these results.

For the Australian landscapes (R. Kaplan & Herbert, 1987*), the perceptual differences between the two samples were more extensive. Using the Australian preference ratings, three categories emerged: Eucalypt Forests, often with strong vertical tree trunks and gray-green foliage (distinctly preferred by the Australians); Australian Scrubland which is quite open and arid; and Pastoral, Grazed, and Relatively Manipulated spacious areas (Fig. 3.13). The relative preference for the three groupings was similar for the two student samples.

Using the American preference ratings yielded five categories. These exemplify the kinds of issues that have consistently emerged in the perceptual analyses that are summarized in chapter 1, on categorization. Openness and smoothness of ground texture are salient properties of the perceptual process. Similarly, vistas afford visual access on a larger scale. Finally, trees play an important role in perception of the natural environment. Thus the combination of tree cover and ground texture distinguishes among several of the categories (Fig. 3.14). The Vista scenes were significantly preferred by the American students. Whether this is a cultural effect or contextual (given this set of scenes, the vistas provided a bigger picture in the context of less familiar settings) requires further research.

The results of these two studies suggest that familiarity with the local natural setting has an effect on perception. The effect of such familiarity on preference is, once again, mixed. Knowledge of the local scene does not assure increased preference, and the novelty of foreign places may enhance it. How much "culture" plays a role in these matters is an open question.

Subcultures

The study of the Australian landscapes (R. Kaplan & Herbert, 1987*) had an added dimension that challenges cultural interpretation of preferences. Another group of Western Australians participated in this study. These were members of a Wildflower Society, sharing an interest in botanical and conservation matters. In exchange for a presentation by Eugene Herbert, then with the Forest Department of Western Australia, they consented to view

Figure 3.13 Examples of each of the Australian-based categories of the Australian landscape (R. Kaplan & Herbert, 1987*). *Top row,* Eucalypt Forests; *middle row,* Scrubland; *bottom row,* Pastoral, Grazed, and Relatively Manipulated settings.

the slides and rate them in terms of preference. Whereas the two student samples, just discussed, permit cross-cultural comparisons, the two Western Australian groups can be assumed to share a culture and to be acquainted with the local natural environment. Nonetheless, these two groups differ in age and education (with more diver-

Figure 3.14 Examples of four of the five American-based categories of the Australian scenes. *Top row, left,* Forest/Forest Vista category, *right,* Open Woodland/Field category. *Bottom row, left,* part of a category labeled Rough-textured, Arid, Wooded, *right,* Vista category. The remaining, unpictured, category was Open, Smooth Texture.

sity in both respects on the part of the Wildflower Society members). Perhaps more important than either age or formal schooling, the two Western Australian groups differ in their interest in the natural environment.

Once again the correlation was high, .81, between the two Australian groups, and the composite mean preferences for all 60 slides were virtually identical. However, for 22 of the scenes the Wildflower Society members indicated significantly greater preference, and for 11 scenes the opposite was true. Thus, for over half the scenes, the two groups were far apart in their appreciation of the local countryside.

The most consistent factor in the differences in preferences between these two samples related to the issue of use of native plants. Pines and pine plantations have been the source of considerable local debate, with some vocal citizens resenting the introduction of exotic species. The Wildflower Society members could be expected to be sensitive to this issue and to prefer virgin forests and scenes reflecting more typical native vegetative associations (and, in fact, they preferred scenes in middle row, Fig. 3.13). The Australian students, by contrast, showed no disfavor of pines (Fig. 3.13, lower right) or of farmland or of other less pristine indigenous bush. These strong differences suggest that familiarity can take several forms. Greater knowledge and concern for the types of species are evident in these preference differences.

The Wildflower Society sample was not large enough to permit a separate CIM analysis, and it is thus impossible to determine whether their perceptual experience is notably different from the students'. The differences in preference are evident, however, when one compares mean ratings for the two Australian groups, using the Australian-based categories. These differences are apparent in the greater distaste for manipulated landscapes and the greater appreciation both for arid, open, coarse-textured views and for scenes of characteristic Western Australian forest.

Different conclusions about similarities and differences in both environmental perception and preference would have resulted if the study had included any two of the three samples. Though both the Australian and the American (student) samples represent "western culture," clear differences exist between the American and Western Australian landscapes. The fact that the two samples that share the same landscape show major differences in preferences has important implications for public participation, for the management of land resources, and for our understanding of subcultures.

Similar subcultural differences have been found in many other contexts as well. Daniel and Boster (1976), for example, included 26 different groups in their validation study of the Scenic Beauty Estimation method. They report generally high agreement with respect to scenic preferences, with the exception of the four groups representing range (or cattle-growing) interests. Porter's (1987) study of preferences for the agricultural countryside also included several discrete public groups, varying in their familiarity with farming and farm scenery. The dairy farmers in the study region, the northwestern part of Washington, showed the most extreme range of preferences, far preferring the open farmland scenes to any other group.

Ethnicity

There are hints in the literature of differences in environmental preference between black and white groups. Across quite a few studies in distinctly different settings, the pattern of these differences is quite consistent. There is great reluctance to attribute these differences to ethnicity, however. In fact, in several instances the results are reported in such a way that it is difficult to determine that ethnic differences are present. Schroeder (1983), for example, describes the groups as differing in "urban" background, with no mention of ethnicity. Zube and Pitt (1981) report distinctly different preferences on the part of a large sample of Virgin Island residents (as compared to Yugoslavian students and various American samples). That the Virgin Islanders are black is true (Pitt, personal communication, 1986), but not mentioned. Given that many of the studies that have found differences between blacks and whites are based on small samples and that it is difficult to determine whether these differences are not also attributable to other factors (e.g., urbanity, age, affluence), it is important to be tentative in drawing conclusions. At the same time, however, it is important to acknowledge that such differences may exist and that this aspect of subcultures is no less likely to lead to preference differences than are the various others.

Washburne and Wall (1980) provide a useful analysis of these issues. They argue that different recreational patterns on the part of American blacks cannot be explained in terms of hardships and "culture of poverty." "The resurgence of interest in ethnic cultural roots . . . supports the existence of true subcultures" (p. 11), and a pluralistic view of American society has replaced the assumption of assimilation as the goal of all minority groups.

Anderson's study (1978*) included black and white high school students and local residents. These participants were not urban. They lived in a rural area where forestry is a major factor in the economy (though at the time of the study unemployment was very high). When he went to the area to take the photographs for his study he consulted some local individuals about where he might find some particularly attractive scenes. At least one black resident explained to him that "there weren't any anymore; the trees are all grown up now." The discomfort with the forests was evident from other comments as well. For example, when asked for advice about getting back into town a black merchant provided some lengthy instructions. When asked whether it would not be shorter to take a different route he replied: "You don't want to go through the forest, do you?" (Anderson, personal communication, 1977).

These are admittedly anecdotal descriptions. Anderson's findings, however, support these impressions. For both students and residents, the white participants significantly preferred scenes of dense forests. The white students were more favorable toward the heavily manipulated (clearcut) areas, though neither student group was fond of these, nor were the adults. Among the residents, blacks expressed much greater fondness for what Anderson called "planned spaciousness," though these were highly preferred settings for all groups. Not only is the tree cover less dense in these scenes, there are also suggestions of human influence, such as roads and picnic tables. (As with previous studies, here again the very high correlations - between .81 and .91 - when comparing residents and students, blacks and whites, would fail to reveal the important differences among these groups.)

The preference pattern Anderson found in a rural context is surprisingly similar to results we obtained in the urban context. Participants in this study (R. Kaplan & Talbot, 1988*) were 97 black residents of three low-to-moderate-income black neighborhoods in Detroit, Michigan. These participants were asked to sort 26 photographs into five piles corresponding to their preferences. The scenes depicted typical outdoor areas, including unmanicured wooded areas, lakes and rivers, landscaped parks, picnic areas, and residential street scenes.

Figure 3.15 shows some of the most preferred photographs in the study (means well over 4.0). These included parks or neighborhood scenes with both natural and built elements. Many had a sidewalk or built pathway as well as park equipment or a picnic shelter or a portion of a building. The ground texture was smooth, and the overall impression was of a well-kept area. Some of the least preferred scenes (means below 3.0) are shown in Figure 3.16.

Figure 3.15 Among the scenes highly preferred by the Detroit residents.

The bottom left scene may look familiar (see Fig. 3.1), as it had been part of the storm drain study (R. Kaplan, 1977a*) discussed earlier. These unfavored scenes depicted unmanaged areas with heavy undergrowth. Tree density, weeds, and a scrubby appearance contributed to these low ratings.

These preferences were also reflected in the participants' perceptions. The CIM-based category Local Parks and Walks groups together scenes depicting paved walks as well as other built elements in the context of nature. The top two scenes in Figure 3.15 are included in this cluster. Both Woods and Path and Woods and Water were far less favored and were similar in including larger and more densely forested scenes (top left and top right in Fig. 3.16, respectively). Pathways and small openings are evident in the former, and a lake or river is in the foreground in the latter grouping. Preference for these two categories was equivalent (providing another example where water is

Figure 3.16. Among the least preferred scenes in the Detroit study.

not highly preferred). These perceptual groupings are similar to results reported by Getz, Karo, and Kielbaso (1982), who also studied black Detroit residents. They found higher preferences for trees on neighborhood streets than for larger wooded areas.

After completion of the sorting task participants were asked to describe aspects of the outdoor areas in the photographs that they particularly liked or disliked. Orderliness, safety, and visibility within an area were the leading concerns reflected in these comments. Trees, built features, and neatness were most mentioned as assets. Though trees are greatly appreciated, the ones in the least liked scene (bottom left, Fig. 3.16) were singled out as negative instances. They were interpreted as looking dead, and the scene as a whole was described as disorderly and cluttered. Heavy undergrowth was associated with the threat of physical danger.

The same scenes and task were used in a pretest prior to the Detroit study. The sample was small, including 21 white and 10 black residents of Ann Arbor, Michigan. The responses of the two black samples (Ann Arbor and Detroit) were quite similar, with only three scenes showing significantly different ratings. By contrast, the differences between the white participants and either of the black samples were substantial. Not only were most ratings significantly different, but the pattern of most and least preferred was reversed in many cases. In fact, the scenes in Figure 3.15 were all rated among the least liked (means between 2.5 and 3.0) by the white participants, and the scenes in Figure 3.16 were all among their highly preferred (means between 3.5 and 4.0).

As Schroeder (1988) has pointed out, some have concluded that blacks may be less interested than white Americans in nature and the outdoors. The results of our research do not support this contention. In fact, the indications are clear and strong that natural settings have great importance to the black Detroit residents. The pattern of findings across various studies, however, suggests that human influence, neatness, and openness are far more vital to some groups than to others.[4]

Age as Subculture

Teens are often identified as a subculture. In certain respects this is just as appropriate for other age brackets as well. It is also just as true with respect to age as it was for other group comparisons that it is at best difficult to ascertain whether findings showing group differences can accurately be attributed to that group designation. For example, if an older group is found to prefer scenes with greater human influence (as was true in Miller's 1984* study), is this difference attributable to the age per se, or are other factors more helpful in understanding the results? Experience with the effects of fluctuating economies might well have been substantially greater for individuals now in their fifties and older than was the case for their age-mates a century ago. It must go without saying that background variables must always be viewed with caution.

Balling and Falk (1982) reported intriguing findings with respect to developmental variation in environmental preference. Their study included four scenes for each of five biomes (savanna, deciduous, coniferous, rain forest, and desert). The participants spanned a wide age range, from students in grades 3, 6, and 9 to college students, "adults," and "14 retired citizens." They describe this sample of 548 participants as "extremely mixed" in terms of other background characteristics. The slides were shown twice (with the order of the task counterbalanced), once to obtain ratings of preference as a place to live and the other time as a place to visit.

Two aspects of the results are particularly noteworthy in the context of this discussion. First, for the two youngest groups (ages 8 and 11) the preference for savanna scenes was significantly greater than for deciduous and coniferous settings, although the latter must be far more familiar to the children. For all the other age groups these three biomes were equivalent in preference. Desert and rain forest scenes were always significantly less liked than the other three habitats. Second, the preferences of the 15-year-olds were consistently lower than those of the adjacent age groups. The plots of preferences for the various biomes all have a remarkable dip at that point in the curve.[5]

Balling and Falk do not discuss this striking preference pattern of the adolescent participants. Zube, Pitt, and Evans (1983) also studied "lifespan developmental" variations in preferences and included younger age groups in their investigation. Unfortunately, they combine 12–18-year-olds into a single age group. This group is not distinct from the 19–35-year-old group in the correlational analyses presented. The report of the study also includes no information about the magnitudes of preferences for the diverse scenes that had been used. It is thus impossible to tell whether the actual preferences for different environments differed across the lifespan.

Medina's (1983*) study sheds some further light on this question of young people and preference. He found the seventh and eighth graders in his study (from predominantly black Detroit, Michigan, schools) to have strongly different preferences than the adults (individuals engaged in environmental education all over the United States). The two groups thus differed in terms of at least four subcultural categories: age, ethnicity, place of residence, and knowledge.

What the students seemed to appreciate most were settings that suggested activity, places to do things. Scenes with buses, for example, suggest that one can go somewhere, and scenes with stores suggest that one can buy something. By contrast, tree-lined streets, preferred by the environmental educators, may be nice to look at but communicate little action. One can interpret the youths' preferences from various perspectives: (1) The preference pattern is similar to the black preferences discussed previously (higher preference for more developed areas). (2) Like Balling and Falk's 15-year-olds, these youths show generally lower preference for more natural settings. (3) The pattern also resembles the findings that people prefer what is most familiar (greater preference for row houses and front-yard lawns than for places dominated by trees). (4) The results also fit a stereotype of the teen subculture (an eagerness to go places and do things, at least in the outdoor context). Whatever the interpretation, the vast preference differences between these young teens and the people who are involved in teaching about the environment must be recognized if environmental education is to achieve its goals.

Summary

Although these studies range widely with respect to human groups and environmental contexts, they show that the balance between nature and human influence is indeed an important factor in understanding preferences. It is easiest to see the subcultural differences as the variant solutions to this theme. For some, nature is favored if it is groomed, orderly, and with indication of human-sign. For others, nature that appears unmanaged and wild is the more preferred. To label only one end of such a suggested continuum as "nature" would undermine the strong positive feelings common to most of these groups with respect to the natural environment.

FORMAL KNOWLEDGE OR EXPERTISE

To a large extent the experiences we have discussed so far have been vicarious or circumstantial. As a resident or visitor or member of a particular group one has encounters with the environment that may be affected by those circumstances. By contrast, the comparisons examined in this section are related to a more intentional pursuit of knowledge about the environment.

Although this is a convenient distinction to draw, it is important to realize that it is only somewhat helpful. The major problem in making the distinction is that learning goes on all the time, and its effect on perception and preference happens without intention or awareness. In chapter 1 we spent quite some time developing the concept of expertise and discussed how experts' perception is different from people who do not share that expertise. In a sense, residents have an expertise about their local environment, and members of a subculture have an expertise with respect to that group. The kinds of expertise we will include in this section, however, are related to professional training.

Preference and Knowledge

We know of relatively few results that point to differences in preference attributable to knowledge about the environment.[6] The wilderness data, showing a decline for preference of swamp scenes after firsthand experience with such settings, provide some indication. However, the knowledge of the swamps obtained from the outing was more physical than cognitive. The different appreciation of the environment exhibited by the Wildflower Society members may or may not be related to formal knowledge about the role of native as opposed to exotic species. There is a strong element of nationalism involved in this issue; one can also favor conservation in principle without extensive knowledge.

A study by Hodgson and Thayer (1980) addressed the possibility that preference is influenced by the suggested interpretation of the scene. Scenes labeled with more "natural" labels (lake, pond, stream bank, and forest growth) were ranked more favorably than when the identical scenes were given labels suggesting greater human influence (reservoir irrigation, road cut, and tree farm). Thus the information provided by the label became part of the interpretation of the scene. Presumably, the labels without any pictures would reflect similar differences in preference.

Gallagher's (1977*) study of prairie restoration at the CUNA facility in Madison, Wisconsin, included a question about how informed the participant was about the restoration and its intent. For the small handful of people for whom he had such information, there was a suggestion of greater preference for the prairie grouping ($p < .10$) on the part of the informed participants.

Buhyoff et al. (1979) studied whether information about beetle damage affected aesthetic judgments. Of the 10 scenes in the study, 8 showed various stages of southern pine beetle crown damage, with colors varying from yellow-orange and red-brown to black, while the remaining two scenes taken along the same stretch of the Blue Ridge Parkway had no such damage (control slides). Four groups participated in the study: introductory forestry students, outdoor recreation students, a nonforestry group including Sierra Club members and residents, and a forestry professional group. For all but the last of these groups, half the participants were told that they would be assessing scenes with southern pine beetle damage, and the rest were told that they were participating in a "landscape preference test." (The professional group was assumed to know that the scenes reflected disease.) Scenes were presented pairwise, and individuals were asked to pick the scene they least liked in each pair.

Of the two control slides, one was ranked most liked regardless of background and information. The other one was ranked second by the informed groups and third by the others. The relative preferences of the scenes showing beetle damage, however, were significantly different for the informed and uninformed groups. For the informed groups, the ranking of preference corresponded very closely to the degree of damaged vegetation in the scene, and there was virtually no difference among these groups, despite their different background knowledge. For the uninformed groups, by contrast, preferences were actually enhanced with increasing amounts of orange-brown foliage since they probably interpreted this as fall coloration rather than the result of infestation.

Keyes's thesis (1984) explored knowledge obtained by virtue of eight interpretive signs along a hiking trail in the Great Smokies. The signs each contained two to three sentences along with an illustrative sketch and referred to information specific to the trail. The signs were mounted in such a way that it was possible to install them on some testing days, permitting direct comparison of preferences between hikers who had the signs available and those who did not.

Keyes reports that of the 298 hikers exposed to the signs 92% indicated that they read at least half of them and that 90% "felt the presence of the signs improved their hiking experience." Eight of the scenes in Keyes's photo-questionnaire were taken where the signs were located (though the photographs , of course, did not include the signs). The preference ratings for six of these eight scenes were significantly greater for the hikers who had signs available (in one instance at $p < .08$). One of the other two scenes was so highly favored by all participants that the sign could hardly increase preference (mean rating of 4.8 for hikers without the sign, where 5 was the top of the scale). Keyes explains that for the remaining scene the sign was actually not immediately in the setting of the photograph. It is particularly noteworthy that the least preferred scene in the study, a view of tangled underbrush, showed a significant increase in preference. The sign in this case specifically addressed "beneficial aspects of tangled underbrush for wildlife."

These studies suggest that preferences can be positively or negatively impacted by interpretation. Not only longer-range experiences but even very newly acquired information can change reactions to the natural environment. Though people in the business of communicating information are committed to the role of knowledge, they are not as keenly aware of the importance of the information in what people like. Yet the preferences are likely to be central in determining what actions the public will take.

Environmental Professionals

There are several questions that are important to consider with respect to the perceptions and preferences of individuals trained in professions that relate to the environment. Given their greater knowledge about the environment, one might expect such people to be more discerning and more differentiated about what they prefer. One might also, however, expect environmental professionals to like natural environments regardless of content or spatial organization because these are the kinds of environments to which they have chosen to devote themselves professionally.

In terms of perception, as discussed in chapter 1, professionals would be likely to categorize the environment in terms of the salient categories of their profession. In other words, the process of training in the respective specialty is intimately involved with learning to see the environment in a particular way. That is not to say, however, that the professional recognizes a change in perception. Each of us assumes, unless something unusual suggests otherwise, that we see the environment similarly to other people. It would be reasonable to assume, then, that experts' perceptions would be different and that they would not be particularly adroit at identifying how nonexperts would categorize the environment in question.

A few studies have addressed aspects of these issues, although there remain many unanswered questions. In most cases the number of professionals included in any one study is very limited, and in some cases students training to become professionals were used as the "professional" group. So, once again, the results are suggestive and certainly worth considering, but the call for further study is loud and clear.

Preference

In addition to students and residents in Anderson's (1978*) study, 27 professionals employed in forestry-related occupations in the study area were also asked to rate the scenes of various forest practices. The professionals' ratings were noteworthy for their strongly positive outlook. For the Heavily Manipulated Landscapes, distinctly disfavored by the local residents and students (means around 2.2), the professionals were on the positive side of neutral (mean 3.4). For the other four groupings their mean ratings were indistinguishable and high (means of 3.8–4.0), whereas the preferences were far more differentiated for the nonprofessional groups. Except for the highly preferred Planned Spacious Openings, the professionals' preference ratings were significantly higher than the residents' in all cases.[7]

Anderson also found that one of the informational predictors played a very different role in accounting for the professionals' preferences. The rated Mystery of the scenes in the study accounted for 7% of the preference prediction for the foresters. For the students and residents, by contrast, Mystery accounted for 39% and 42%, respectively.

In Medina's (1983*) study the professionals' preferences were more differentiated. Compared to the junior high school students, the environmental educators differed significantly on seven of the eight CIM-based groupings. They preferred the scenes showing more natural areas, with lower residential density, and open spaces that were more enclosed. The students, by contrast, preferred scenes that reflected a more urban context, with fewer trees and greater opportunity for travel. The two groups showed no difference in preference for the least preferred grouping, industrial scenes.

Studies oriented to design-related professionals have also shown differential preference effects. In contrast to studies reporting high correlations between preference ratings by different groups, Buhyoff, Wellman, Harvey, and Fraser (1978) reported close to zero correlation between landscape architects and lay groups. Once again, such results may obscure some interesting patterns among the ratings.

A study that included architecture and landscape architecture students ("designers") as well as students in an upperclass psychology course ("clients") found significant group effects in environmental preference (R. Kaplan, 1973b*). Using CIM procedures, three categories were identified. For the Nature grouping (wooded and relatively enclosed settings), the architecture students were far less favorable than were the other two groups. This pattern was reversed, however, for the grouping emphasizing Building Complexes (generally graphically rendered, architecturally striking scenes). The scenes in the third category, Part Buildings with Nature, included both a portion of a building and relatively open areas with landscaping that provided a context for the visible portion of the building. For this category, the landscape architecture students were significantly more positive than the architecture students, who were more favorable than the client group.

In addition to the preference ratings, participants were asked to rate the scenes in terms of Coherence and Mystery. The results showed strong differences in the degree to which these informational predictors affect preference for the three groups. Using partial correlations (i.e., looking at the effect of two variables while ruling out the influence of the third), the role of Coherence in predicting preference was strong (.89) for the architecture students and less powerful for the other two groups (.72 in each case). The role of Mystery, by contrast, was very strong for the client group (.93) and less strong for the architecture students (.66). The landscape architecture students were almost equally affected by Coherence (.72) and Mystery (.80). For all three groups, the two predictor variables were negatively related to each other (partial r's between −.32 for the landscape architects and −.58 for the students with no design training). Along these lines, it is also interesting that the architecture students rated the Nature scenes as significantly lower with respect to Coherence than did the other two groups, whereas for the clients the Part Buildings grouping was rated as much lower in Mystery than it was for the designers.

Summary

With respect to preferences there is indication that professionals differ from lay groups. Whether the professionals are more or less differentiating in their appreciation of the natural environment is not clear. The studies that suggest that fewer distinctions are made have involved forestry-related professionals. Studies that have involved design professionals and environmental educators suggest that these experts are quite differentiated in their preference judgments.

Although based on very meager findings, the suggestion that the experts weigh the role of informational aspects of the setting differently than do lay groups is important to acknowledge. It helps make preference differences seem less idiosyncratic and may make it easier to reach a common understanding when experts and the public struggle to reach a mutually satisfactory solution.

Perception

Several studies have explored the ability of professionals to understand or express the preferences of the public. The Buhyoff et al. (1978) study, which found no relationship in personal preferences between landscape architects and non-design-trained students, also explored the question of perception. The landscape architects (students and faculty) were given the "free responses" produced by the students, reflecting "any factors which affected their preferences for the landscapes." Based on these, they were to anticipate the students' preferences. How well they accomplished this task depended on whether they had professional experience in design. Perhaps contrary to expectation, the rankings by the subgroup with such professional experience were unrelated to the actual student preferences; the rankings by the landscape architects who had no actual design experience, however, correlated .86 with the student client group's rankings.

Anderson (1978*) asked six foresters to categorize the scenes in his study in terms of forest practice or land uses. The results of this simple request were quite complex. Collectively, they identified 16 categories, which Anderson organized into five broad themes: plantations, developed campground, reforestation and harvest practices, roads, and openings. The categories that emerged based on preference ratings were distinctly different. Roads did not appear as a separate grouping, and reforestation and harvest practices were reflected in two separate categories. What might be considered an "openings" cluster also included the developed campground scenes, thus combining two of the foresters' themes. Red Pine Plantations is thus the only category with direct correspondence between the preference-based and forester-identified categories. Anderson had not asked the foresters to anticipate the groupings of the empirical findings, and it is, therefore, reasonable that the two sets are not identical. Even so, the lack of agreement among professionals comes as a surprise.

Hudspeth (1982*) and Miller (1984*) asked the professionals in their studies to group the scenes to reflect their expectation of the public's categorization. Once again, relatively few professionals were invited to do the task, and the agreement among them was surprisingly low. In both studies, the professionals accurately anticipated a category based on "industrial" scenes and correctly expected these to be relatively low in preference. Some other categories that were nominated by more than half of the nine "key influentials" in Hudspeth's study did not correspond to the results based on the Category-Identifying Methodology (for instance, public access, the Barge Canal [which empirically combined with other undeveloped areas], and views of lakes and mountains). In Miller's study six of the nine planners and resource managers included the natural bald spots with "undeveloped" scenes, whereas empirically these combined with clearcuts.

When the data from all the professionals are pooled, in any one of these studies, there is moderate correspondence to the empirically derived "citizen"-based results. The match between any one expert's prediction and the preference-based categories, however, is consistently poor. Given that professionals do not necessarily have the opportunity to pool their insights in many situations where their decisions affect the public's visual resources, these results give one pause. The planners, foresters, resource managers, and designers in these various studies demonstrate that their judgments do not correspond well to the public's and that these experts tend to be unaware of the difference. In fact, both Hudspeth and Miller reported that the experts found the task very difficult. They had to be reminded on more than one occasion to separate their own values from their expectations of the public.

These results provide strong support for incorporating the preferences of the local public in assessment of the local visual resource. Schauman (1988a, 1988b) provides a model for such a procedure for the agricultural countryside. It is equally appropriate and needed in other contexts as well.

DOMINANT AND VARIANT THEMES

The results summarized in this chapter make one wish there were more: more research with large enough samples to validate the effects based on the groups included here as well as more studies to explore differences based on many other comparisons. At one level these findings lead to more questions than solutions. One cannot be sure of their generalizability to other settings and contexts and to other groups. In other respects, however, it is hardly surprising that perceptions of and preferences for the natural environment are not universal. Findings of differences are important to recognize and must be taken into account in planning, policy, and management decisions. Rather than pointing to "complete idiosyncrasy," that one simply cannot account for taste, such findings suggest that there are differences but they are not idiosyncratic.

Each of us has experienced changes in our own reactions as we have gained familiarity with and understanding of the environment. Things that were initially bewildering or "far out" became commonplace and comfortable.

And things that seemed comfortable became tired and uninspired. But some old familiar things became no less distasteful, and other old familiar things remained very endearing. The differences among the various groups we have discussed in this chapter, whether based on home range, age, expertise, or other subcultural designation, reflect similar patterns. Familiarity per se does not fully explain the preferences; on the other hand, familiarity necessarily interacts with people's reaction to the environment.

The ability to understand the environment must necessarily be a dynamic process. Making sense of things is contextual and varies with constantly changing information. Places that seemed safe may suddenly be seen in a completely different light because of newly acquired information. A nearby weedy field one might have complained about can suddenly look different when one hears that it might be converted to a parking structure. Similarly, the eagerness for involvement is also likely not to remain static. There are times when exploration is exciting, and there are times when one wants to be sure the ground beneath one's feet is firm and safe with as little excitement as possible.

What the variant results suggest, then, is that the preference framework developed in chapter 2 is still useful and powerful. However, the relative weighting of its components is likely to depend on many factors. For example, the relatively greater emphasis on the understanding side of the matrix, reflected in higher preference for scenes that are orderly and controlled, may reflect one or more of several concerns. It may be especially salient when other aspects of one's life are less orderly so that one desires at least some arenas that make sense. Or one may seek orderliness in settings that one understands less well. With respect to the interior of one's home, for example, one may prefer much greater complexity than for natural areas with which one is less comfortable. The desire for order may also reflect a shared pattern of a culture or subculture. Similarly, the relatively heavier weighting on the side of exploration may also be a function of combinations of various factors.

Despite all the variations, there remain substantial constancies. The strongest of these is the importance of nature itself. The differences among groups have reflected concerns for safety, for order, for apparent human influence, for adventure, for preservation, and for many other qualities. They have not, however, reflected that nature does not matter. Trees and water, flowers and green things, the sense that the plants grow and that they will always be there – these do indeed seem to be as close to universals as one can find. The kinds of satisfactions people derive from such green experiences are the subject of part II.

NOTES

1. Some exceptions to this pattern have been cited. Schroeder (1988) discusses some urban/rural differences with respect to urban park and recreation patterns. Miller (1984*) also reported some preference differences related to this source of familiarity.
2. Schauman's proposed system for visual resource management of the agricultural countryside, developed for use by the Soil Conservation System, takes an important step in this respect by incorporating participation at several stages of the process as well as using "modifiers" to reflect local elements (Schauman, 1986; Schauman & Pfender, 1982). A field test of this procedure is presented in Schauman (1988a).
3. The Australian students were the same for the two studies, with the order of presentation of the slide sets counterbalanced. Different samples of American students participated in the two studies.
4. Several other researchers have reported findings that show patterns similar to those reported here. See, for example, Dwyer, Hutchinson, and Wendling (1981), Peterson (1977), Washburne and Wall (1980), and Zube, Pitt, and Anderson (1975).
5. Balling and Falk (1982) emphasize that preference declines with age. However, their oldest age group included only 14 individuals. The 8-year-olds were high in preference for each biome, and the 15-year-olds were low. For the college-age group and the 35-year-olds, however, there was no decline.
6. A study that examined the effect of orientation, prior to a visit to an urban forest, showed some interesting relationships between learning and preference, although the methodology was different from the studies described here (R. Kaplan, 1976). Board games based on aerial photographs were used to familiarize the children with the spatial layout of the area prior to exploring it on a field trip. The area had a heavy tree cover and was potentially challenging as far as way-finding was concerned. A control group played similar games, but their game board was oriented to a different area, one they did not visit. The group that was oriented to the environment that they did visit enjoyed the field experience more, was more enthusiastic about taking further trips, and

indicated a greater likelihood of returning to this area in the future.

7. Balling and Falk (1982) also include foresters in their study and report significantly higher preference ratings by this group. The foresters, however, were compared to all age groups and not to their age-mates. Their pattern appears from the plots to be in much the same range as the adult group's.

PART II

BENEFITS AND SATISFACTIONS

NATURAL materials and likenesses of natural scenes are pervasive elements of our civilization. Is it all a conspiracy? The florists see to it that arrangements bedeck major events, that we say it with flowers, whatever "it" may be. Picture postcards in thousands of drugstores depict classical scenes of woods, gardens, and beaches that could be found anywhere, but the inscription names a cherished local landmark. And big, expensive books, displayed on many coffee tables, frequently feature glossy photographs of nature: droplets on a sunlit leaf, majestic trees, fern fronds unfolding. Are these simply efforts to ensnare the unsuspecting, to force upon the public yet another product in exchange for hard-earned dollars?

Although it is clear that a considerable industry profits from flowers and other aspects of the natural environment, there is reason to suspect that materialism alone cannot explain the pervasiveness of such products and that other pecuniary explanations are far from the whole story.

After all, wreaths preceded any current commercialism by a good many centuries, and little children have needed no prompting from the florist industry to gather bouquets of long-stemmed dandelions for their mothers. One can wonder how much of the news that is communicated on familial long-distance phone calls concerns the current blossoms in the garden and how may slide shows of faraway journeys include scenes of colorful markets and pushcarts of posies.

The widespread participation in nature-related experiences and activities is, from some perspectives, peculiar indeed. Why do people bother growing flowers anyway? This is often a costly, frustrating, time-consuming activity that yields nothing useful and simply adds work. Why do people hike and backpack? Why spend money on equipment and travel in order to get hungry, lost, bitten by bugs, and risk getting sore feet? And why go somewhere that you have already seen before or to a place you have learned about through others' stories or pictures? There is, after all, the widely accepted dictum that humans are rational animals, that they judge all possible outcomes and choose the ones that maximize their gain, that economic considerations are overriding in human motivation and decisions. An examination of the benefits and satisfactions people derive from plants and nature encounters provides a heavy dose of evidence contrary to such a rationalistic position. The purpose of part II is to explore some of these benefits and satisfactions.

In part I the discussion was not specific as to the setting. Part II, by contrast, is divided between natural settings that are some distance away and those that are in home range. These categories appear at first to be similar to the extremes of a continuum that has become central to recreation planning by federal agencies, such as the U.S. Forest Service. The Recreation Opportunity Spectrum (ROS) has become widely accepted as a tool for classifying, inventorying, and mapping recreation resources based on settings, activities, and experiences (Cordell & Hendee, 1982). It identifies six levels of "naturalness": primitive, semiprimitive nonmotorized, semiprimitive motorized, roaded natural, rural, and urban.

For the ROS, however, "nature" at its wildest and "nature" when proximal to population centers bear little resemblance to each other. The more primitive the setting, the more people can experience solitude, tranquility, self-reliance, and closeness to nature. At the other extreme, however, "experiencing natural environments, having challenges and risks afforded by the natural environment and the use of outdoor skills are relatively unimportant." Instead, people are described as seeking more passive and affiliative experiences at the "urban" end of the spectrum (Driver & Brown, 1978).

The material in chapter 4 does not contradict the ROS characterization of the more primitive setting. The studies in chapter 5, however, suggest that the role of nature in the urban context is in many respects remarkably similar, rather than, as the ROS model implies, diametrically opposed. Not only are many similar experiences sought in the nearby-natural context, but many similar benefits seem to be available as well.

It is often suggested that younger, better-educated, and

more affluent people are the ones who avail themselves of wilderness experiences. The distant, pristine, and seemingly endless tracts of land with rock outcroppings, virgin forests, and perhaps some cool, clear streams are, in fact, for some people the only settings that can be called "natural." To spend time in such primeval settings restores one's depleted mental energies. Writers such as Thoreau, Muir, and Olson have written eloquently of the joys and benefits of such wilderness encounters.

The research discussed in the next two chapters shows that wilderness is not the only setting for experiencing such restorative experiences. The satisfactions achievable in the wilderness require neither the pristine natural setting nor that one be educated and affluent. The studies draw on a variety of research tools, including structured and open-ended questions and ratings of photographs. Though the participants are far less articulate than the well-known naturalist-writers, the data nonetheless corroborate many of the naturalists' insights.

Chapter 4 focuses on distant natural environments. The chapter deals with a research program that lasted over a decade in which individuals who participated in a wilderness program were also willing to participate in various empirical studies. The research looked at such issues as self-esteem and changes in self-concept, the process of becoming familiar with the environment, and the time element both as one accepts the pace of nature and as one reenters the everyday world. The results of these studies provide some insights into factors that enhance the restoration of the human spirit.

Chapter 5 focuses on the nearby, everyday natural environment. Several of the studies on which the chapter is based look at the role of nature in the residential context. The view from the window turns out to have a strong bearing on satisfaction with one's neighborhood. Whether that view includes trees and fields and other natural elements thus makes a substantial difference in people's feelings about where they live. Knowing that the natural resource is available, as opposed to direct use of it, seems to be particularly important. The chapter also deals with perhaps the most prevalent form of nature involvement, gardening. The satisfactions people derive from this activity are clearly pertinent regardless of demographic distinctions. Gardeners derive much more than vegetables and flowers from their efforts.

CHAPTER 4

A WILDERNESS LABORATORY

As the stereotype goes, the scientist is clad in a white lab coat. The laboratory is a microcosm where the scientist can exercise maximal control. There the scientist can test the effect of one event on another, thus confirming the previously articulated hypotheses.

This chapter is about a quest that lacked any of these features of science. But like much of science, stereotypes notwithstanding, it was a process of guided discovery, where theory is refined by data and data gathering is motivated by theory.

The laboratory was some 17,000 acres of land, much of it densely forested. The "scientist" was nowhere in sight. Control was at the mercy of climate and the vicissitudes of a particular group of strangers brought together for a fortnight in this apparently endless, largely trackless piece of wilderness. Nor were there clear hypotheses to be confirmed. There were, however, many questions, and the questions changed in the course of this 10-year adventure.

There are many extraordinary aspects to the research that was part of the Outdoor Challenge Program, the focus of this chapter. The very fact that a program such as this had a research component is in itself most unusual. In the early 1970s some individuals involved with the U.S. Forest Service's North Central Forest Experiment Station saw the wisdom of studying the benefits of a program like Outward Bound. Their modest support continued for more than 10 years, permitting the issues that were explored to change as our knowledge accumulated. This support meant extremely reasonable rates for the participants in exchange, as it were, for their willing participation in the research program.

The Outdoor Challenge Program had been initiated by Robert A. Hanson, who remained its director throughout those years. Not only was he essential to the running of the program and to procuring the funding for it, but he was an eager and willing partner in the research program as well. That meant that there was a continuous interactive relationship between the program and the research effort. The program benefited from the research results, and the research benefited from the opportunity to implement new ideas based on previous results. To have such a process continue for a 10-year period is a researcher's dream.[1]

But to say that the Outdoor Challenge Program is unique in its long-standing research orientation is not to say that this provided an ideal setting for doing research. The number of uncontrollable variables was great; they serve well to remind one that this is a slice of "the real world," albeit one that few people have the chance to be part of. To be able to obtain meaningful results despite the "noise" is all the more encouraging and instructive. One must be mindful, however, of the limitations of the situation in exploring the insights gained from the program.

THE OUTDOOR CHALLENGE PROGRAM

The Outdoor Challenge Program involved backpacking through a large wilderness area in and around the McCormick Experimental Forest in Michigan's Upper Peninsula. During the first 8 years of the research collaboration, the trips were 2 weeks long; for the last 2 years, 1980 and 1981, they were shortened to 9 days.

Although there were some variations in the program over the years, many features remained constant. The research aspect was always evident in the materials the participants were asked to complete (before setting out on the first day, at the end of the program, at various times in between, and even after returning home). In addition, participants were provided with journals and asked to write their feelings and whatever reactions they might have to their experience on the trip.

Other constants were the physical demands of the program. Much of the hiking and backpacking was through dense areas (sometimes so dense one could not see through the vegetation), swampy regions, and some steep country, much of it without trails. A practice hike the first evening motivated instruction in map reading and compass orientation. During the next several days participants took turns orienteering and leading on the hikes. These went through dense, largely trackless forest for the first several days until the group reached the McCormick tract where the land is more open and there are occasional old logging trails. One further constant feature of the programs was a 48-hour solo, a time when each participant was alone at a lake, generally not in seeing or hearing distance of others. During the 2-week programs, following the solo the leaders departed and the participants hiked on their own to one of the other lakes in the area and then returned to base camp.

Three types of groups participated in the program: high-school-age males, females of the same age, and coed groups consisting of individuals past high school age. Some of the adult groups included high school students (in which case the group was all of one gender). Group sizes varied widely, between 3 and 12, with at least two leaders accompanying each group. These leaders were generally local outdoorsmen or individuals from the county mental health clinic.

Most of the participants were from Michigan's Upper Peninsula, especially from Marquette and nearby small communities. Others came from Ann Arbor as well as from the Detroit metropolitan area. The students were recruited from local high schools where a film was shown in the spring describing the program. Many participants inquired about the program because of announcements, brochures, and contacts with previous participants. The adults included public school teachers from the Marquette area as well as college students or recent graduates, housewives, office workers, and so on. Table 4.1 summarizes information about the number of groups and participants during each of the years of the research collaboration.

The Outdoor Challenge Program was originally modeled on Outward Bound programs. Over the years the emphasis shifted away from survival skills and physical hardship. A rappelling component that had been incorporated for boys' and girls' groups (not for adults because the time of their outings presented staffing problems with respect to the rappelling program) was dropped in 1979, and the intent of solo was shifted as well. Some of these changes are best described after we look at the intent of the Outward Bound model in the next section.

Throughout its existence, Outdoor Challenge was concerned with the environment and with the participants' un-

Table 4.1. *Outdoor Challenge participants*

Year	No. groups	Boys	Girls	Adults	Total
1972	2	10			10
1973	2	12	8		20
1974	1	8			8
1975	3	13	6	4	23
1976	3	6	12	8	26
1977	2	6	3	3	12
1978	3	6	7	6	19
1979	3	5	3	3	11
1980	6	11	11	13	35
1981	2		3	9	12
Total	27	77	53	46	176

derstanding of how to survive comfortably in this environment: how to make one's way through the woods, cope with physical discomforts, and work through one's own fears. Increasingly over the years, the program also placed continuing emphasis on the opportunity for individual reflection. The research itself has contributed to this theme, as participants were repeatedly asked to describe their reactions to their surroundings and to their daily experiences.

CHALLENGE PROGRAMS

Outward Bound provided a model for many challenge-based programs. These programs quickly teach participants " skills that enable them to survive in primitive environments and that expand their horizons in terms of their own self-expectations" (Hanson, 1973). Hanson, then Director of Psychological Services at a community mental health center in Michigan's Upper Peninsula, describes the attraction of this kind of program: "These programs, combining high interest with inherent motivation, are especially appealing to the mental health professional, and are a hopeful offering to a society struggling with alienation and distrust."

Burton (1981) estimated that over 300 wilderness programs of this genre were available at the time of his thorough review of the literature. Many of these are oriented toward special populations including juvenile delinquents, psychiatric patients, corporate managers, and educators. Such programs are offered in many parts of the world. In fact, Burton traces the origins of Outward Bound to Kurt Hahn, who founded two schools that served as precursors to Outward Bound – the first in Germany, his native country, and the second in England when he was forced to flee Germany in Hitler's time. The first Outward Bound school, however, was started in Wales in 1941 "to train merchant seamen to withstand the rigors of wartime life" (Burton, 1981). The success of the Aberdovey School led in the course of the next 20 years to the founding of similar schools in Europe, Africa, Australia, and North America. Their emphases varied to some degree, with risk, endurance, adventure, and group effort prominent features.

Nold (cited by Burton) considered Outward Bound "as an experience of self-discovery and enhancement of self-esteem that uses the challenges found in the natural setting as a medium." In reviewing the literature on the benefits of such programs, Burton considered five critical features to be (1) a group of at least four participants, (2) a leader or instructor, (3) "a challenging set of problem solving tasks," (4) a "contrasting or novel physical environment," and (5) a duration of at least 7 days.

Though such programs have been popular and their benefits have been touted, the research efforts to document the benefits are less voluminous. Nonetheless, Burton found some 73 research references, which he summarized. Like others before him, however, he found much of the research seriously flawed in terms of various methodological criteria. Driver, Nash, and Haas (1987) mention the following among the prevalent deficiencies in such studies: small sample sizes, lack of appropriate control group, lack of follow-up studies, reliance on subjective reports and nonstandardized measures, and insufficiency of statistical tests. With these flaws in mind, Burton's 73 studies were reduced to 19 that he considered "valid studies."

Burton's conclusions, based on either the larger set of studies or the more restrictive one, were that the evidence supports a certain amount of positive gain and that there is a surprising lack of negative findings. The gains are largely in "self-concept," and they seem to be more evident with delinquent adolescent populations than with "normal" adults. Although many "outcome variables" have been studied, there is no apparent, consistent gain in such characteristics as grade point average, "behavior," self-actualization, or self-awareness.

THE FIRST PHASE OF OUTDOOR CHALLENGE RESEARCH

In its early phases, the Outdoor Challenge Program fit comfortably in the range of programs reviewed by Burton. The earliest research efforts also support the general trend of his findings. The first reported findings (R. Kaplan, 1974) were based on a very small sample, 10 adolescent boys who had participated in two groups, and a control group of 25, matched in terms of age, gender, and region. As with studies reported by Burton, there was no question that the Outdoor Challenge participants gained substantially with respect to many aspects of woodsmanship skills (Fig. 4.1). Compared with the control group, whose differences over a 6-month period were minimal, program participants also showed gains in certain aspects of self-concept. They were more realistic with respect to their

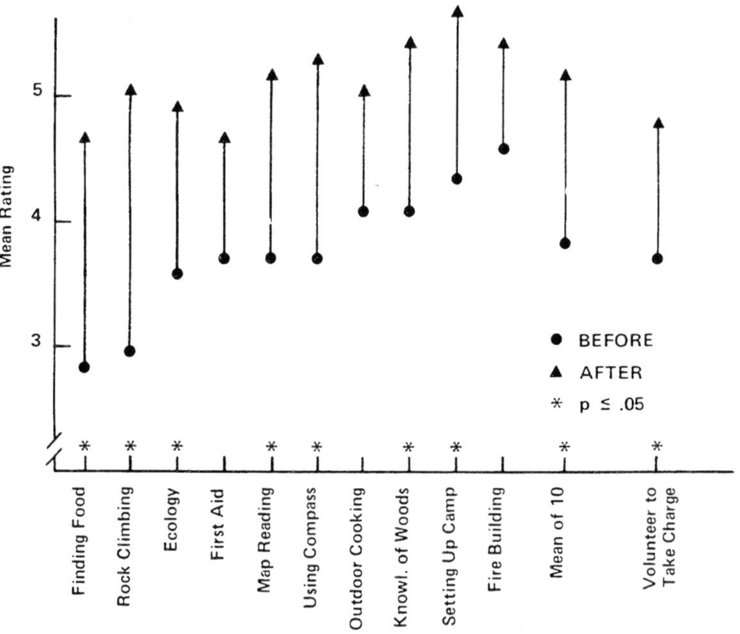

Figure 4.1 Mean rating for specific woodsmanship skills, both before and after wilderness experience (from R. Kaplan, 1974).

own strengths and weaknesses, felt they had greater self-sufficiency in the use of their time and talents, showed a greater sense of concern for other people, and had a more positive view of themselves.

Comparisons of responses to the 6-month follow-up survey showed two interesting differences between these two groups of adolescent boys. In answer to "If you could change yourself in any way, in what way would that be?," more than a third of the controls mentioned physical characteristics - being taller, handsomer, "a little more bigger." Only one of the Outdoor Challenge people mentioned such qualities, and his was the desire for improved eyesight! The other striking difference was in the fact that about half of the Outdoor Challenge group, and only one of the controls, answered this question by stating they would not want to be different. One fellow was quite adamant about this: "I know I'm not perfect, no one is. I'm content to stay the way I am and keep all the imperfections" (R. Kaplan, 1974).

The following year (data collected in 1973), the research broadened in scope. A total of 267 high school students, both male and female, completed an initial questionnaire, and 200 of these returned the second questionnaire some 6 months later. Table 4.2 indicates the composition of this sample. In addition to Outdoor Challenge participants, individuals participating in other backpacking and camp programs were included. There was also a substantial con-

Table 4.2. *Participants in 1973 study*

	Spring			Fall		
	M	F	All	M	F	All
UP control	85	52	137	64	41	105
AA control	13	17	30	7	12	19
All control	98	69	167	71	53	124
Outdoor Challenge	12	8	20	12	8	20
Isle Royal	8	10	18	2	9	11
Wind River	5	5	10	4	5	9
Camp program	18	26	44	10	20	30
Mtn. Field	2	6	8	2	4	6
All programs	45	55	100	30	46	76
Total sample	143	124	267	101	99	200

trol group drawn from the same schools and classes in the Upper Peninsula (UP) and in Ann Arbor (AA).

The groups were by no means equivalent with respect to a large set of measurements prior to the summer. With respect to self-esteem and woodsmanship skills, however, there were no significant differences. The material completed by the participants in November showed many interesting contrasts. (A summary of the measures used in the study is in Appendix C.)

Not surprisingly, the Outdoor Challenge group indicated a profound and highly significant improvement with respect to virtually every one of the woodsmanship skills. The exceptions were, once again, skills not included in the program (e.g., canoeing and outdoor cooking). Compared to the other two backpacking-oriented groups in the study, the Outdoor Challenge participants indicated a higher skill level on seven of the items. These other two groups, however, indicated significantly greater proficiency in 8 of the 12 items than did the "camp" group, a 5½-week program where participants "work as a community exploring and learning new skills, developing environmental awareness and sensitivity to the natural areas." Though based on self-reports, these results correspond closely to expectations. The emphasis of the Outdoor Challenge program is precisely on woodsmanship or wilderness skills, and the other backpacking groups necessarily used such skills to a far greater extent than did the camp program, which did not depend on the out-of-doors nearly as much. The groups did not differ initially, nor did they have knowledge of each other's ratings in responding to the mailed fall questionnaires.

To see whether such self-perceived competence is related to self-esteem, we compared those with relatively high overall woodsmanship skill scores with those with lower scores – separately for each subsample. Even for the Outdoor Challenge group, where everyone scored relatively high on these skills, the higher scorers reflected a more positive self-esteem. For each of the other subsamples, differences in woodsmanship skills were related to the Realistic Task Orientation component of self-esteem.

Our findings with respect to self-esteem correspond to Burton's conclusions. There were some gains and some "no changes," and no negative findings. Participants in any of the programs who had shown initially low scores on the Positive View scale were twice as likely to show a gain

6 months later than were the controls. Of the Outdoor Challenge participants who had initially reflected a high Negative View of themselves, half no longer reflected such a self-view (others showed no change on these items).

The question concerning how one would want to change reflected findings similar to those in the previous year. Once again, the Outdoor Challenge participants mentioned physical changes significantly less often than the controls. The same was true for the other two backpacking groups but not for the camp group. In other words, the groups with the greatest improvement in specific skills related to the out-of-doors were least inclined to mention a desire to change in physical respects. Both the camp program participants and the Outdoor Challenge participants, however, reflected a desire to change in the direction of greater independence, self-discipline, patience, and self-reliance. About 40% of each of these two groups, who did not mention these aspects in the spring, did in the fall (few in the other groups mentioned these at either time).

These studies avoid some of the methodological pitfalls that have plagued research on challenge programs, but they nonetheless have their share of flaws. The results are based on self-reports and to a large extent do not use standardized measures. The sample sizes are not impressive, and there was attrition in the fall responses. Caution must clearly be exercised in making inferences about the effects of the summer experience, given many other factors involved. In viewing each result in isolation, it is easy to raise alternative explanations and reservations. The overall pattern of results, however, seems to make such a conservative interpretation unduly restrictive. The mutually supportive pattern of results and the pervasiveness of the observed changes strongly suggest that the different summer experiences led to changes that were not only substantial but also sufficiently durable in character to be evident some months later (see R. Kaplan, 1977b, for fuller description).

SOME QUOTES FROM PARTICIPANTS

Silence is a funny thing. I don't hear it often.
Last night I think I experienced the most I ever have.

It's quiet and very peaceful
I am damn glad to be here
It's not scary at all

good night
ps. the bugs are not bad

I thought I just herda bare but it was a fly
The silence was a terrific new experience

I was surprised that solo meant so much. I thought that I would be the same about the woods and thinking about myself but I'm really different about myself now.

I don't understand it I just feel so much alive
I want to yell and scream and tell everybody

When I go home I know I will want to tell my friends about this experience. I will become frustrated and bitchy because either I won't have the words or they won't have the ears. Whereas now I am happy.

The thing I think about most about the program is that I can see so much more beauty in the earth in its natural state than I ever could before.

The richness of the comments in the journals, the answers to the more structured questions, and the nature of the participants' observations led to changes in the research emphasis after the initial two years. Rather than focus on the comparison of program participants with control groups, we focused on what was happening within the program itself: the processes that unfold during the two weeks and the nature of the changes that participants experience.

The next three main sections of this chapter explore some of these shifts in research emphasis. The first, "Getting to Know," examines the process of becoming familiar with a setting that initially seems alien and even frightening. We were surprised to find that the process is quite rapid. The second section looks at the solo experience and what happens in an enforced period of solitude. The third research topic focuses on the changes that become evident as one returns to what had been one's "normal" life. It is not unusual for people returning from the wilderness to refer to this as "reentry," suggesting that there is a marked transition. In fact, this is a time of confronting many contradictory feelings.

GETTING TO KNOW

The initial research phase convinced us that "something important was happening." At the start of the trip these youths generally expressed some apprehensions and felt ignorant about the ways of the woods; they emerged 2 weeks later feeling comfortable with the setting and themselves. Thus, in the course of a mere 2 weeks, these adolescent boys and girls were becoming proficient in some skills and were viewing themselves differently. Beginning with the 1974 participants, the research addressed the question of "getting to know."

This research concern was no secret. On the very first day of the program participants were handed questionnaires and were introduced to our interest: "We would like to have your help in figuring out what kinds of things help one feel at home in an area like this." Five topics were included in these questions: how well they felt they *knew* different aspects of the setting, how much each of 10 physical characteristics helped provide a *sense of place,* how misleading or *confusing* each of these seemed, some possible sources of *worry,* and how much certain things contributed to being *comfortable.*

Ideally, one would track these issues repeatedly, being careful to obtain measurements at the same point in the program for all the groups. In the first year we approached this task by using open-ended questions and asking participants to record their observations frequently. This led to a more structured format the next year, based on the early replies. Since it was apparent that the process occurred rather rapidly, in the second year we requested information at four points during the trip. In subsequent years the questionnaires were shortened and were used only three times because it was evident just how quickly most changes were occurring. The ideal of having carefully timed responses was not realizable in the context of this field laboratory. Nonetheless, the results discussed here are based on material that was collected at three points that were roughly equivalent: initially (late the first day or during the second day), soon thereafter (generally the day after the initial data point), and on the sixth or seventh day when the group reached its base camp within the McCormick tract. Eight groups and 61 participants, in the course of 4 years, are represented in these analyses. (All ratings used 5-point scales, with a higher value indicating a greater amount of the rated quality.)

Knowing the Setting

For each of these items dramatic changes in ratings occurred between the first and second data points (Table 4.3).

Table 4.3. *Knowing the surroundings*

	Knowledge rating			Change (p-level)	
	C1 (1st campsite)	C2 (2nd campsite)	W2 (2nd week)	C1–C2	C2–W2
Layout of the land	2.7	3.6	3.9	.001	.001
Other group members	2.7	3.6	4.2	.001	.001
Plants and wildlife	3.1	3.4	3.6	.05	.05
Water, food sources	3.4	3.7	4.2	.01	.001
Compass	3.4	4.1	4.2	.001	—
Map	3.8	4.3	4.3	.001	—

The two items dealing with individuals' estimates of their ability to use the most essential tools in this environment – the map and the compass – showed no further changes. The more general items, however, those dealing with understanding one's social and physical surroundings, continued to increase through the first week of the program.

Sense of Place

The list of features included the following: stars and sky, islands, rock formations, ravines, roads, hills, streams, lakes, trails, and swamps. Only the last two of these showed any significant changes in features that help provide a sense of place. Trails were rated significantly higher at the last data point, reflecting the lack of trails encountered early in the program and the fact that the remnants of the Bentley Trail were important features. The other significant change, swamps, is particularly noteworthy. It is unlikely that many people who have not experienced this program would feel that swamps can make one feel at home!

Sources of Confusion

Participants were asked to consider the same 10 items in the "sense of place" question with respect to their being "misleading or a source of confusion." For the group as a whole, none of the features became significantly more confusing. Swamps and ravines were the only items to show a change between the two initial points, and islands and streams were found to be less confusing as more time elapsed. By the second week, islands and lakes were by far the least confusing features of the environment (mean ratings below 2.0). Swamps, by contrast, maintained their standing as among the most confusing (with hills and ravines) of the physical features, with ratings around 2.7.

During the initial year of this study, when participants provided open-ended responses, their entries indicated that, in order to be helpful in way-finding, features need to be unique and dependable. Uniqueness, however, changes with experience. What may seem a landmark early on can become commonplace after similar configurations are seen repeatedly. Thus rock formations and cliffs seemed more promising as landmarks on the first day than later on. Lakes, however, seemed to maintain an element of uniqueness. The sun and compass bearings, by contrast, remain dependable - even when one becomes disoriented in a swampy area or on an often disappearing trail.

Sources of Worry

As would be expected, the worries tended to decrease as the days went by (Table 4.4). The only exception to this was with respect to being away from home, where the mean rating rose after the initial day and then leveled off. Of all the "worry" items, however, this one received by far the lowest rating, averaging only 1.5 by the second week. In fact, none of the items received high ratings. The weather and bugs maintained their position as somewhat greater worries.

Table 4.4. *Sources of worry*

	Worry rating			Change (p-level)	
	C1 (1st campsite)	C2 (2nd campsite)	W2 (2nd week)	C1–C2	C2–W2
Away from home	1.3	1.4	1.5	.05	
Animals	1.9	1.6	1.6	.01	
Getting sick	2.3	2.3	2.1		.02
Being alone	2.3	2.2	2.0		
Getting hurt	2.5	2.6	2.3		.05
Sore and tired	2.5	2.7	2.1		.001
Solo	2.5	2.3	2.4	.05	
Getting lost	2.5	2.3	2.0	.05	.01
Weather	2.6	2.7	2.6		
Bugs	2.7	2.4	2.3	.05	

The other worry showing a relatively higher initial level, getting lost, was the only one to decline significantly at both sampling points. Much of the skill acquisition central to the Outdoor Challenge Program is related to reducing the likelihood of getting lost. With this fear becoming less salient, other worries can also dissipate. Dangerous animals, for example, become less problematic. This is presumably not because of numerous encounters with bears and a growing sense of accomplishment that might come from successfully fending against such creatures. More likely it is a fringe benefit of the overall reduction in fearfulness that is an outcome of the increased understanding of one's surroundings.

The concern about being "sore and tired" is less cognitive. Here the drop did not come immediately. This is partly because of the particularly strenuous aspects of the first days of the program; by the time a week had gone by the participants were both more accustomed to these aches and presumably in better physical condition. In addition, the pace had slackened somewhat by this point in the trip.

Sources of Comfort

The striking aspect of these ratings was not so much the changes (there were few) but the features that were found to be particularly comforting. Even at the initial sampling point two items received very high ratings (over 4.0): the "peace and quiet of the woods" and "feeling that nothing is in a big hurry." These high ratings were true for males and females, teens and adults, initially and at subsequent sampling points.

Two additional items became sources of comfort after the initial day, both achieving ratings of 4.0 by the second sampling point: using a compass and "noticing things, hearing sounds of nature."

Conclusions and Implications

It is easy to take for granted the things that one knows. In fact, it is difficult to appreciate what it is that one knows and what role such knowledge plays. The time when one most readily appreciates the role of experience is when one lacks it. When needed knowledge is missing its importance can loom large; and yet, when one has it, it so readily seems trivial. Such is the way of expertise!

For the novice wilderness explorer the lack of knowledge is a matter of some urgency. It seems that the number of things one should know are endless and that the consequences of one's ignorance could be fatal. Prior to the outing participants desired to know survival techniques, all the flora and fauna of the region, the geography, the weather, and so on. In a matter of a few days, however, there is sufficient experience that it seems there are really very few things one needs to know. It is all quite matter-of-fact and routine. The participants repeatedly voiced the feeling that one does not need to have or to know a lot in the woods. Personal attitude is said to be the important thing: the need to keep cool and stay calm, to have confidence.

What the veteran participant has that was missing even a few long days earlier is some imagery and vividness as well as knowledge motivated by circumstances. Information that might otherwise have been considered "academic" is suddenly central and relevant. The knowledge that the prevalent flora consists of second-growth northern hardwoods with some patches of hemlock, as well as stands of mixed paper birch, aspen, and balsam fir, is not very useful for someone with minimal background in the out-of-doors. Once in the midst of such an area, with evidence of earlier logging pointed out, the concreteness of this information is greatly enhanced. Suddenly there is a desire to know. But once the knowledge is gained it is difficult to appreciate the role it plays.

The experience in the setting, the imagery one has gained, makes it possible to develop a model, an internal map of that environment. As with all internal maps of the many things people are knowledgeable about, their existence and utility are often invisible to their owners. It becomes clearest that these maps are in our heads when there is a mismatch. Without such internal organization of our knowledge it would be meaningless to be surprised. To realize that something is unlikely in a particular setting requires experience with what is likely. Thus the notion that all that is needed in the face of potential unpredictability is the "personal attitude of staying calm" is the expression of an internal model that can guide one to "keep one's cool."

Such a theoretical perspective did not come out of this research. It guided the research to begin with. The results of the research support this perspective in broad outline, but there were also a number of new insights and even some surprises.

Perhaps the least expected and most striking finding of this research is the rate at which the getting-to-know process takes place. In most areas, expertise is much slower to develop. This extremely rapid rate of change in people's sense of comfort and their confidence deserves closer examination. It may be the consequence of the total immersion that characterizes the program. It may also be related to the heavy emphasis on orientation skills right from the start. Before starting the program participants indicate that what they would want to know about a place is, for all intents and purposes, "everything." Asked a similar question some months after the program, the focus is much more on the terrain. Comparably, for the adult women especially, greater confidence in their ability to read the map and use the compass is central to their sense of comfort in the setting. Feeling comfortable with orientation provides an important way to counter other fears. Fear of snakes and bears is difficult to dispel in a direct way, but these fears subside as confidence in one's own skills increases.

In surprising ways these results support the discussion of categorization and expertise in chapter 1. The experience with the environment changes us quickly and quietly. By and large it is not a process to which words are attached. Nor are people aware of how radically affected they are by the way they see the world.

SOLO AND SOLITUDE

The "solo" has been a feature of many challenge programs. Outdoor Challenge, like other Outward Bound-type programs, began by making solo a relatively arduous experience. Participants were provided with minimal equipment – a piece of clear plastic for shelter, a hunting knife, some salt, and a survival cup for cooking. Whether to fast or to forage was the individual's choice; whether to sleep or keep a fire going was also left to the individual. In fact, all decisions had to be faced without the support of others. Faced with oneself and the sounds of the natural world around one, the solo was rarely a neutral event.

Based on participants' comments and journal entries, a new approach to solo gradually evolved. It appeared that the minimal provisions led more to thoughts and concerns about "making it" than to spending the time on reflection and on noticing the environment. In later years of the program participants were provided with a sleeping bag and minimal food. In addition, exercises and discussion prior to solo were oriented toward becoming aware of one's surroundings and coming to terms with being alone.

During the years of the getting-to-know project, participants were also asked a series of questions relating to solo, both before and after the experience. Apprehensions were relatively low prior to the solo; participants expected it to be a time to think about themselves and their futures. After the solo, however, they indicated the experience was somewhat less challenging than they had expected, more boring than anticipated, and not quite as "productive" in working out things about themselves as they had hoped. Nonetheless, solo was a time for "being in tune with nature" and for "thinking about the future and what's really important."

Though the adults, boys, and girls did not differ significantly on any of the items before solo, after solo there were some differences. The solitude aspect of the experience seemed particularly satisfying to the adults. As a group, the boys reflected a less positive impression of the experience. They found solo to be more boring, less comfortable, and more worrisome than did the others. In fact, of the nine individuals who did not complete solo over the years of the program, all but two were boys; generally, these were younger boys (around age 15) who had expressed a preference for activities with high risk. For people like that there is too little going on within the confines of the solo setting; they are not yet mature enough to be sensitive to the fascination inherent in a whole little world. Their expectations were for something much more exciting. The opportunity to contemplate and be in tune with one's surroundings is not the sort of "activity" to which they had envisioned devoting 48 hours.

Solo is certainly not a "guaranteed peak experience." One woman commented that it was the hardest thing she had ever done, including childbearing. Several other adult women indicated that they found it depressing and debilitating. On the other hand, the assessment of the experience even a few days later (on the last day of the program) already showed some changes, even for individuals who had rated solo as less positive. "Having made it" became an important aspect of the individual's self-assessment.

In the last 2 years of the program, when it had been shortened to 9 days, a 48-hour solo was still maintained. The shorter program was necessarily more intense, with about the same amount of hiking prior to solo but in a shorter time period. The opportunity to be alone, to "hear the silence," may thus have been even more welcome. This is at least a possible interpretation for the more positive reactions by these participants to some of the items. The Solitude scale, in particular, was more highly rated in these 2 years (mean 3.93) than previously (3.45). This included such items as "just being alone, the silence, peaceful, not lonely, a great experience," rated on a 5-point scale.

Analysis of the responses to the questions about solo (using the same Category-Identifying Methodology discussed in chapter 1) resulted in another aspect of the experience that emerged as separate from the Solitude dimension. It included items such as "figuring out what kinds of things are important to you" and "thinking about who you are and who you want to be." We called this Reflection. For this category there were no differences in scores as a function of trip duration. In fact, the ratings on Reflection were consistently high, with means around 3.9.[2]

Reflection has not been the subject of much discussion, but solitude has received some attention in the recreation literature. The Wilderness Act of 1964 identifies solitude as a distinguishing characteristic of wilderness, although there is no accepted understanding of what constitutes this desired state (Hammitt, 1982). Lucas (1980) found that the most frequently mentioned reasons for choosing a roadless wilderness experience were its "wilderness qualities," including solitude, isolation, and no motors.

Assuming that people prefer minimal encounters with others, solitude has often been interpreted to mean physical isolation. Research on "carrying capacity" has focused on the effects of encounters while in wilderness and backcountry areas. In many cases the findings do not support the notion that people seek total isolation or that encounters per se detract significantly from the experience (Cordell & Hendee, 1982). The size of the group encountered, whether it is at camp or on the trail, and the expectations with respect to the setting itself are all factors in the satisfaction with the wilderness experience (Stankey, 1980). In his comparison of nine wilderness areas, Lucas (1980) found that solitude was mentioned as a factor affecting the satisfactoriness of the experience by only 10–15% of the participants. The environment itself and scenery, by contrast, were mentioned by 42–62% of the respondents.

Because most people who spend time in the wilderness are not alone but in small groups, solitude must necessarily be interpreted in terms of the experience of the immediate group with respect to encountering other such groups. The solo experience, by contrast, concerns a truly solitary experience. Whether seeing (or hearing) others in that context has a positive or negative effect may be quite different. Outdoor Challenge participants were asked about these issues, and their responses are somewhat surprising.

Table 4.5 summarizes the results for the participants of the last 2 years (N = 47). Though most people did not see anyone, for those who did the experience was not necessarily negative. Hearing human voices, by contrast, violated the spirit of solo for some, whereas it was reassuring to others. The traces left by previous people were the most detracting of the visual aspects, although many felt neutral about this or even found them a positive factor. Hearing birds was considered a positive experience in most cases. Hearing other animals, by contrast, was disturbing to six

Table 4.5. *Effects of sights and sounds on solo*

	Negative	Neutral	Positive	Not applicable
See				
People on solo	3	1	11	32
Other people	1	4	7	35
Use by others	12	18	9	8
Birds			44	3
Other animals		1	35	11
Animal tracks	1	4	20	22
Other animal sign	4	3	23	17
Hear				
People on solo	7	3	9	28
Other people	1	3	6	37
Birds		6	41	
Other animals	6	2	33	6
Chainsaw	3	3	3	38
Airplane	17	19	11	

of the participants. Even in relatively remote areas the sounds of airplanes are not uncommon. Each participant heard at least one airplane. No one rated this as "very disturbing" although about one-third found it negative. For this one item the difference between the adults and adolescents was significant. The adults minded this intrusion far more than did the youths.

Conclusions and Implications

It is not unusual to be alone. Some people like to have the radio turned up at such times whereas others prefer quiet. Some are reassured to see signs of other people whereas others prefer to feel alone with themselves. What makes solo a very different experience of being with oneself is the context. One is alone not in a house or tent or even on a bench in a park but in a natural setting with minimal evidence of human influence. One is alone with the creatures who share that world, with the plants that grow there, and with one's thoughts. One cannot rely on the radio or a tape deck or the phone or the refrigerator or even a book to create diversion.

That such a situation would be uncomfortable and even distressing would hardly be surprising. It is far more surprising that the effects of 48 hours of such solitude are generally so positive. Under normal circumstances, people rarely notice their natural surroundings in any detail. They are probably relatively insensitive to their own thought patterns as well. The contrast presented by the solo experience is striking in both respects. At the same time, for many participants the solo is by no means a uniform high. It is quite characteristic for the low point to come as the first night approaches and for spirits to be raised in the course of the second day. The feelings upon rising after the second night are particularly positive. The inner sense of peacefulness takes time to develop.

CHANGES IN SELF-CONCEPT AND REENTRY

People seek a backcountry experience for many reasons. There is a substantial literature on this subject that shows some consistencies across settings and activities and also shows great diversity in motivational and preference profiles. As Knopf (1983) indicates, "there are now literally scores of in-depth analyses of outcomes desired from recreation experiences in natural environments." He goes on to say that "themes of stress mediation, competence building, and the search for environmental diversity dominate the literature."

Driver et al. (1987) show comparisons of recreationists at 15 settings, varying to some degree in their wilderness designation, in terms of their "experience preference domains." At each site between 80 and 1,567 people were sampled, for a total of 5,360. Of the 16 domains presented, "enjoy nature" heads the list in all but three cases, and it is in second place for two of those. The next most important domains are "physical fitness," "reduce tensions," "escape noise/crowds," and "outdoor learning." Considerably less important in these ratings is "introspection/spiritual"; and aspects of "achievement/stimulation" (e.g., reinforcing self-confidence, competence testing, seeking excitement) also receive more moderate ratings.

Outdoor Challenge participants have not been asked questions that permit direct comparison with these analyses. If they had, their answers might well have been quite similar. These categories, however, fail to capture much of the effect of these experiences. Categories based on participants' responses to questionnaire items and journal entries suggest a very different structure. Data from the last 2 years of the program, completed both before and immediately after the trip, show significant changes in moods and feelings with respect to several domains. (The grouping of the items in Table 4.6 is based on the CIM procedure. Here again a 5-point rating scale was used with 5 = very much.)

For many of the participants the extended immersion of this backcountry area leads to some surprising new perceptions of both the environment and themselves. S. Kaplan and Talbot (1983, p. 178) describe these new relations as evident in the journal entries by the end of the first week: "The wilderness inspires feelings of awe and wonder, and one's intimate contact with this environment leads to thoughts about spiritual meanings and eternal processes. Individuals feel better acquainted with their own thoughts and feelings, and they feel 'different' in some way – calmer, at peace with themselves, 'more beautiful on the inside and unstifled.'"

Table 4.6. *Moods and feelings (1980–81 participants, N = 47) (R. Kaplan, 1984)*

Scale items	Coefficient Alpha	Mean rating Before	Mean rating After	t	p
Psychological energy					
Self-confident, generally in control of things, able to concentrate, relaxed, physically active, full of energy	.81	3.71	4.24	4.76	.0001
Simple life style					
Responsible in accepting work chores, curious about things in general, enjoys getting along on less, eager for ways to simplify life, enjoys dealing with problems	.69	3.69	3.92	2.78	.01
Positive outlook					
All's right in the world, content with life as a whole, realistic about oneself, feels great to be alive	.68	3.55	4.07	5.59	.0001
Tuned in					
In tune with nature, happy with a slow pace	.53	3.52	4.08	4.49	.0001
Hassled					
Readily irritated by other people, easily irritated at small things, harried, hurried and rushed, caught up in the "rat race"	.77	2.24	1.84	3.26	.005

Reentry

The multiplicity of these changes, the vivid realization that "there is more to the out-of-doors than trees," hits many of the participants when they suddenly find themselves removed from this newfound home base and return to "civilization." For some the sudden change to a much faster pace as they leave the area by car already triggers strong reactions. For many the reentry to their regular routines and family setting requires readjustment. The journals we asked them to keep during the first period after their return are rich in images both of enjoying some of the luxuries of life (one's bed, different food, etc.) and of the tribulations of contending with people, noise, a faster pace, and lack of time for the cultivation of inner peace. Many noted the ugliness and artificiality in their surroundings, the unnecessary urgency in their activities, and the superficiality in their friendships.

Despite the change in trip length to 9 days, participants during the last 2 years showed very similar reactions to the reentry process. In earlier years as well as the last two, although the return had many negative aspects, the comments were largely about positive memories. Remembering the wilderness as awesome and compelling, recalling the tranquility of the woods, noticing nature in the everyday environment, and a feeling of self-confidence were often-repeated themes in the journals.

Reading the comments the participants recorded during their first week back, one is struck by the innumerable contradictions that are expressed in the reentry experiences. But although these may be seen as wildly fluctuating moods or even a total loss of coherence, they are also reasonable dilemmas as one tries to assimilate new perceptions into old realities.

- Being more patient and more irritable

  The irritability tends to be in the face of irrelevancy, of a pace that is too fast; the greater patience concerns what the person has come to decide really matters. Perhaps, in combination, these are a result of greater clarity about what is and what is not important.

- Being more able to concentrate and being distractable

  The attentional capacity has had a good rest and, from a theoretical perspective, one might expect a greater capacity to concentrate. The distractability, by contrast, may be related to the clearer sense of priorities, making it more difficult to attend to things that seem less interesting and important. A further source of distraction is a tendency to mind wandering, as participants think back to their recent experience.

- Feeling good about the world and feeling terrible about the world

This is, alas, an all too appropriate reaction. Having experienced a deep sense of tranquility, having discovered some important new insights about oneself, perhaps having even had a feeling of being at one with the surrounding natural world, one has every reason to feel good about things. On the other hand, one cannot fail to realize that one's experience was neither characteristic of what goes on in the world nor easy to repeat very often. The world one is returning to is, in many ways, opposite to the world one is coming from. Distractions abound, people's purposes conflict, and the natural world is neither dominant nor, for that matter, widely appreciated for its potential benefits.

- Feeling inner peace and feeling a sense of alienation
- Feeling tranquil and feeling harried
- Being less fearful of danger and more "on guard"

These and other similar contrasts mentioned by the participants are understandable on the same terms as the previous examples. There is an inner world of harmony and peace but at the same time an outer world that is perhaps now seen as potentially more undermining of what one values than was previously the case.

- Experiencing joy in simplicity and enjoying having one's luxuries

Both sides of this coin can be viewed as benefits of the experience. On the one hand people become aware of the advantages of simplicity and are better able to appreciate it. On the other hand they also appreciate certain aspects of civilization that are easy to take for granted.[3] Both of these feelings can contribute to a sense of renewal, a sense of being able to see more that is positive in one's everyday experience.

INSIGHTS FROM THE WILDERNESS LABORATORY

When we were approached by the Forest Service in the early 1970s and asked to help "evaluate" the benefits of the Outdoor Challenge Program, then going on somewhere in Michigan's Upper Peninsula, we had little reason to anticipate the benefits that would emerge. The initial studies included comparison groups and follow-up data and tried to assess changes in self-esteem in a variety of ways. The results were, in many ways, comparable to the outcomes of the numerous studies summarized by Burton (1981) in his dissertation.

In the course of the research efforts in subsequent years, we began to see much more fully that the effects of the program were substantial and multidimensional. That is not to say that all participants gained equally or that the changes will endure indefinitely. Nor, for that matter, can we say that such changes are directly attributable to specific aspects of the program. After all, the program changed continuously: Drought and downpours had significant effects on routes and conditions, leaders differed in styles and approaches, and the dynamics among participants necessarily varied from year to year. In addition, the emphasis on survival and endurance changed in response to insights from data, and the testing instruments themselves were modified as we learned from previous years. Nonetheless, the comparability of results from one year to the next was surprisingly high, giving one some confidence that there is a central core of commonality in the program and that it has a significant impact on the participants.

People participating in the Outdoor Challenge Program were exposed to an extended nature experience. The duration in a predominantly natural environment was far greater than what most modern humans experience most of the time. But perhaps even more important than the duration of the experience was its intensity. Nature provided not merely the background for the activities but the content itself. Participants learned to find their way in the terrain, to carry what they needed for survival, to be viable members of the natural community.

There are many aspects in which the world of the Outdoor Challenge participants is a greatly simplified one. The life-style dictated by backpacking in the wilderness necessarily eliminates many facets of modern life. The social environment is also simplified by the unity of the small group's purposes.

The many aspects of simplicity - unity of purpose, lessening of distraction, emphasis on the basics of survival - have an interesting impact on people. Many participants experienced a sense of "wholeness" or "oneness." There is less conflict between what one wants to do and what needs to be done and less that seems arbitrary or irrelevant.

Having such a feeling of wholeness is, not surprisingly,

an exhilarating experience. It provides a sense of well-being, of being renewed, of being restored. It also offers a way of life that permits the recovery of aspects of mental functioning that had become less effective through overuse.

The role of the natural environment is inherent to these experiences. Not only did participants notice more aspects of that environment, but they came to realize that they lived differently and felt differently during their immersion in this setting. The coexistence with other creatures and growing things gave them a new perspective on themselves. The existence of the wilderness became a comforting thought.

Self-discovery

There are a number of factors in a program of this kind that tend to foster self-discovery. These can be divided into two main categories. The first comprises experiences that are out of the ordinary and provide new information to the individual. Thus the discovery of the important role nature can play is an outcome of the special circumstances created by the program. The experience of coping with the physical and mental challenges of backpacking and of solo is also unique to the situation. Some participants feared these activities were beyond their capacity; having met these challenges gave them a new perspective on themselves.

Another sort of challenge was provided by the potential dangers that were thought to lurk in the woods. Over time this fear lessened. There was a growing sense of competence, of being able to take care of oneself. This led to a reduction in the feeling of helplessness, a sense that, if one could conquer one's fear of something as dangerous as the woods, one need not be fearful of anything else.

The other way in which the program tended to enhance self-discovery was through the encouragement it provided for reflection. In a setting less harassing and less distracting than the everyday environment, there was considerably greater opportunity to be open to information that might otherwise be ignored. The situation itself also provided the participants with many new insights. They were coming to see themselves as meeters of challenges, as appreciators of nature, of simplicity, and even of silence.

The process of reflection and self-discovery takes on different meanings for the youths in their teens and for the somewhat older group. For some of the adolescents the program seems to provide a coming-of-age experience. Not only is it as difficult as one could ask for; it also is perceived as important. For many of the adults the experience of something they regard as important and valuable leads to thoughts of how much their everyday life lacks these qualities. Similarly, the experience of tranquility and a sense of peace is in sharp contrast to their daily routines. Such realizations lead to serious reflection on what is worthwhile and what is not. There is also some anxiety about whether these new insights and new resolves can be transplanted into the everyday world when they return.

CONCLUDING COMMENTS

In some ways the findings of such studies of the wilderness experience might seem far removed from reality and of questionable practical value to most of us. Does it take 2 weeks to achieve these gains? Does it require a wilderness setting? Are there shortcuts to achieving a sense of tranquility and to feeling one is a part of the natural world?

First, it is important to acknowledge the problems of defining *wilderness*. As Robinson (1975) pointed out, "The core of the difficulty lies in the fact that wilderness is not merely a condition of the land, but also a condition of the mind evoked by the land." What is legislated as wilderness must have great extent and must be minimally affected by human forces; what is perceived as wilderness need not meet these constraints.[4]

Although our research in the wilderness made us aware of some of the benefits it provides and some of the qualities of becoming "restored," it is unlikely that these are available only in that context. In fact, chapter 5 is devoted to a discussion of various studies that suggest that nature much closer to home provides many parallels.

Thoreau wrote of nature as a source of spiritual renewal and inspiration. A surprising outcome of the wilderness research has been the remarkable depth of such spiritual impacts. This is surprising because of the strikingly different context, much shorter duration, and nonintellectual nature of the participants in our studies compared with Thoreau's experience. It has also been surprising to discover how rarely impacts such as these are discussed in the psychological literature. The quest for tranquility, peace, and silence resonates with what in religious contexts might be considered serenity. Similarly, the sense of oneness is more likely to appear in a spiritual context than in research on human functioning. A third dimension that comes out strongly in these results, the notion of wholeness or what Mary Midgley calls "integration," may be related to the achievement of a coherent sense of oneself. We had not expected to find the wilderness experience so powerful or so pervasive in its impact. Nor had we anticipated that this research program would provide us with an education in the ways of human nature. We have been introduced to some deeply felt human concerns that broadened our conception of human motivations and priorities.

NOTES

1. In addition to the close collaboration with Robert Hanson, the Outdoor Challenge Program depended in many ways on the efforts of Janet Frey Talbot. For most of the years of the program she was a key link to the data, holding major responsibility for interpretation of the journals and for data management and analysis. The discussion here draws on much of this work and on papers she copublished with us. Appendix C provides a list of the major papers on which this chapter depends as well as copies of some of the instruments that were developed for this research.
2. Hammitt and Brown (1984), in a study on the perceived functions of wilderness privacy, also found a clear separation between Solitude and Reflection, though their study was not in a wilderness setting and used different items. Nonetheless, among their reported factors was one reflecting "Emotional Release" and a separate one, which they called "Reflective Thought." They characterize the former as related to the release of physical tension, whereas the latter concerns "releasing psychological stress."
3. Brickman and Campbell (1971) provide a stimulating analysis of phenomena of this kind in the context of adaptation-level theory. They point out that the experiences and activities that once gave pleasure can over time become invisible and unappreciated. The resulting yearning for new and greater sources of pleasure can create a continuing pressure that is hard for a culture to fulfill. Periods of deprivation or at least contrast to the usual pattern provide a means by which culture can reset the adaptation level. Thus, after a fast day even quite ordinary food can be experienced as thoroughly enjoyable.
4. S. Kaplan and Talbot (1983, p. 199) suggest "a psychologically oriented definition of what wilderness must be," consist-

ing of three components: (1) There is a dominance of the natural; (2) there is a relative absence of civilized resources for coping with nature; and (3) there is a relative absence of demands on one's behavior that are artificially generated or human-imposed.

CHAPTER 5

NEARBY NATURE

THE juxtaposition of *nearby* and *nature* strikes some people as a contradiction. The word *nature* is often reserved for areas that have been unaffected by human influence, that have trees and other vegetation, and that have considerable extent.[1] What is nearby for most people, most of the time, could hardly be described as lacking human influence and is unlikely to be vast. Yet vegetation could well be present, and perhaps that feature in itself qualifies for the "nature" designation, even if it is at one's doorstep.

The issues here are not simply semantic. The failure to recognize the satisfactions and benefits that the nearby-natural setting can offer has important consequences. It means that all too often landscaping is considered merely an optional "amenity." Having green things nearby is undeniably pleasant but is often deemed less essential than all that is subsumed by "infrastructure." Noteworthy architectural monuments rise in the cityscape, but funds run out before the landscaping plan can be put into effect. Public housing projects can often be spotted quickly by the total lack of nearby vegetation. The possibility for gardening is all too rarely afforded residents who do not own a single-family home.

Underlying such failure to acknowledge the nearby natural are issues of how one categorizes the natural environment (the concern of chapter 1). Knopf (1987) discusses various bases for such taxonomies. Some of these are based on the social transactions that they facilitate. Others focus on such dimensions as active/passive, control/noncontrol, and participant/spectator. A classification system that is now widely adopted by federal land agencies in this country reflects these considerations. The Recreation Opportunity Spectrum (ROS) designates six distinct types of settings, ranging from primitive to urban. It provides a useful contrast to the studies we will look at in this chapter.

The ROS model includes characterizations of the setting and of the experience. For their "urban" category, they provide the following descriptions:

Setting characterization: Area is characterized by a substantially urbanized environment, although the background may have natural-appearing elements. Renewable resource modification and utilization practices are to enhance specific recreation activities. Vegetative cover is often exotic and manicured. Sights and sounds of humans, on-site, are predominant. Large numbers of users can be expected, both on-site and in nearby areas. Facilities for highly intensified motor use and parking are available with forms of mass transit often available to carry people throughout the site.

Experience characterization: Probability for experiencing affiliation with individuals and groups is prevalent, as is the convenience of sites and opportunities. Experiencing natural environments, having challenges and risks afforded by the natural environment, and the use of outdoor skills are relatively unimportant. Opportunities for competitive and spectator sports and for passive uses of highly human-influenced parks and open spaces are common. (Clark & Stankey, 1979)

A great deal of research led to the ROS framework. Relatively little of it, however, focused on the nearby-natural settings that people see, pass through, and even create for themselves. Nor has this research generally acknowledged "undesignated" settings or activities that are not labeled "recreational." These relatively neglected domains are the focus of this chapter. Urban parks are included in the nearby natural, but so are street trees and backyards, fields and unused lots, courtyards and landscaped areas. Looking out the window is rarely included under "recreation," yet it can constitute an important opportunity for experiencing natural elements. Nor is gardening generally included in "outdoor recreation," yet it is perhaps the most widely practiced "sport."

This chapter is based on a series of studies (described in Appendix D and marked by * in the text) that explored a variety of topics related to the nearby-natural environment. In particular, we address such issues as the importance of size or extent of a natural area, the effect of different degrees of proximity, and the different ways in which a natural area is used. Across these different themes there is an emphasis on the perceived satisfactions and benefits derived from nature settings and activities - especially gardening. Given the different kinds of settings and activities that we call upon here, and the specific issues that these studies address, it is hardly surprising that our conclusions about the quality of nearby-nature experiences bear little resemblance to the ROS "urban" category. Though it is certainly true that nature that is nearby can be used as a social setting, it can also be "a place apart," a setting where tranquility is possible even in the midst of the urban bustle.

THE ROLE OF EXTENT

An important component of wilderness is that it has extent. Nature that is nearby is rarely vast. How essential is size to the benefits derivable from natural settings? In other words, we have seen that the wilderness setting can lead to some strong positive consequences (chapter 4), and it is useful to explore whether the size or extent of the natural setting is essential for the benefits.

The bigger-is-better outlook seems to be prevalent with respect to parks and open spaces, as it has been in many other contexts. But downsizing has become more fashionable, and "small is beautiful" is a more accepted perspective. Relatively little research has addressed the issue of whether size is related to preferences for different natural settings.

In a study that considered this question (Talbot & R. Kaplan, 1986*), we found that participants were fairly accurate in judging the relative sizes of urban nature areas (in this case varying between 0.1 and 18 acres) that were presented photographically. The preferences for the settings, however, were unrelated to the actual or perceived sizes. The participants' answer to questions regarding factors that make an area seem larger and those that enhance their preference were surprisingly similar, however. In both instances features that were frequently mentioned included "spacious, wide-open areas," "trees, especially large and numerous," and "trails and pathways." It is easier to see that the first two of these features would seem to enhance both preference and the perceived size of a setting; why "trails and pathways" should seem to increase size is less clear.

Subsequent studies (Bardwell, 1985*; Talbot, Bardwell, & Kaplan, 1987*) explored the issue of size and preference across a wider range of outdoor areas, and with respect to natural areas with which the participants were relatively familiar. In these studies there also was no relationship between size and preference for those areas that were relatively large. For the immediately adjacent front and back yards at these townhouses, however, the relationship between size and preference was strongly negative (Fig. 5.1). At one site especially, the developer considered it appropriate to provide virtually no front yards in favor of more common space. Although this project won a Sensible Growth Design and Planning Merit Award, the residents' evaluation would certatinly be different.

In some cases there is indication that bigger is definitely not better. Some residents in the previously mentioned studies, for example, considered the adjacent open areas too large to be comfortable (Fig. 5.2). Similarly, results of a study of Detroit black residents (Talbot & Kaplan, 1984*) suggested some fear of large undeveloped spaces and concern about attackers who might be hiding there. They preferred smaller, well-manicured settings, which could more easily be monitored. Washburne and Wall (1980) report similar preferences, based on a national survey, for smaller, nearby parks and for more facilities rather than more land.

Though not in the context of parks, the findings with respect to the availability and views of nearby-natural settings were very similar in a study of multiple-family hous-

Figure 5.1 The front and back yards were considered far too small by the residents in these housing developments (Talbot, Bardwell, & Kaplan, 1987*).

Figure 5.2 Big, anonymous open areas such as these were frequently considered too large (Talbot, Bardwell, and Kaplan, 1987*).

ing complexes (R. Kaplan, 1985a*). Here neighborhood satisfaction was found to be far greater when residents could see even a few trees than when their view was of large open spaces.

One other context that offers some insight into the importance of size is the vest-pocket urban park. In the absence of hard data, one must assume that the popularity of some of these parks (e.g., Paley and Greenacre in New York City) is an indication that their smallness is not detrimental. Our study of a newly built, very small downtown park suggests that a park does not need to be large to be highly valued, and creating one large space was less preferred than creating a setting with many smaller regions (R. Kaplan, 1980).

These studies offer a beginning, but the question of the role of the size of nearby-natural areas remains largely unanswered. One reasonable conclusion is that size needs to be considered in the context of other issues. For example, size and familiarity may be interrelated such that fear of the unfamiliar may lead to preference for smaller, more knowable areas. Size also may be related to the kind of natural area one is considering. Thus, one's immediate and most personal outdoor area may involve different criteria than a more distant, shared area. Finally, there is some indication that, rather than size itself being the important issue, it may be the perception of extent that is of greater significance. An area that seems to have extent, that suggests that there is more to explore than is immediately apparent, has a special attraction. At the same time, intimate spaces are desirable even within a larger area. Thus, rather than focusing on size, it is important to consider how the space is designed to achieve these valued qualities.

PROXIMITY AND DIFFERENT USES

Though size may not be of primary importance in the context of nearby nature, proximity seems to be essential. One of the "patterns" that Alexander, Ishikawa, and Silverstein (1977, p. 305) describe is called Accessible Green. The pattern states: "People need green open places to go to; when they are close they use them. But if the greens are more than 3 minutes away, the distance overwhelms the need." Proximity, from this perspective, is measured in minutes, and these are intended to be minutes spent on foot (as opposed to in a car or on a bicycle).

Proximity could also be measured in terms of physical distance or in terms of perceived distance. Even if a nature place is, in fact, only minutes away, if the distance *seems* to be substantial, the setting is pragmatically far away. Thus a green place that requires crossing a major highway with no traffic light in sight is appropriately considered far away.

Alexander et al. suggest that use – walking to the green area – is an important outcome of proximity. Clearly, distant places are less likely to be used on a frequent basis, but it is less clear that proximal places are, in fact, frequently used. We have found that satisfaction can be more closely related to the perceived availability of a setting than to its use (R. Kaplan, 1985a*). Furthermore, some of the areas for which residents expressed the highest preference were those that they frequented least often (Bardwell, 1985*; Talbot et al., 1987*).

Use, however, needs to be considered from a variety of perspectives. In the outdoor recreation literature, users are people involved in activities. Many sports, for example, take place in an outdoor context, although they are not generally intrinsically involved with the nature setting. It is thus not surprising that the availability of parks and recreation fields and the frequency of sports activities made minimal impacts on satisfaction levels (Frey, 1981*, R. Kaplan, 1985a*).

Other activities bring one in closer contact with nature – walking or hiking, picnicking, gardening. These all involve intentional, purposive activities. Still other uses of nearby-natural settings are more "circumstantial" (Bardwell, 1985*). For example, much of the encounter with the front or back yard may not be to nurture the plants (intentional) but to go to the mailbox or walk to the bus stop (circumstantial). Even for such encounters, the setting can make a difference.

The experience of the natural environment, however, includes other forms of involvement as well. Even what might be considered an active form of involvement can entail a range of intensities. Thus a neighborhood walk certainly calls for greater physical exertion than sunbathing on a balcony. But neighborhood walks can vary in how much physical exertion is involved. When walking for exercise, for example, one might experience the same setting quite differently than when the walk is frequently interrupted to appreciate the early signs of spring.

It is thus evident that *observing* is an important form of involvement with nature (R. Kaplan, 1984a*). Noticing the buds and blossoms, the changing colors, the nest of a bird or wasp are all, in a sense, "uses" of nature. To call these "passive" seems unhelpful. One can observe while pursuing an active use, and one can be passive with or without noticing what is happening. The confusing nature of the conventional terminology in this area should not, however, be permitted to obscure the central issue. Much of the pleasure that people derive from nature comes from such occasions to observe. And much of the observation occurs when people are not necessarily in the natural setting itself but looking from a window. Cooper-Marcus and Sarkissian (1986) point out that the primary basis for judgments of the attractiveness of one's neighborhood is what can be seen from the window of one's home.

It is hardly surprising that studies on windowless settings (including schools, hospitals, work environments) suggest that they are unpreferred (Verderber, 1986). The view of natural areas has been shown to make a difference with respect to health measures (Moore, 1981; Ulrich, 1984; Verderber, 1986; West, 1986) as well as satisfaction (discussed further in the next section, "The Importance of Nearby Nature"). Looking out the window provides an opportunity to let the mind wander. Almost a third of the participants in a study carried out in Detroit (Talbot & Kaplan, 1984*) indicated that nature was important for just such thinking – whether one is in a natural setting or looking at it.

Another form of involvement with the natural setting is even more problematic to consider as a "use." It is related to one's knowledge and imagination, even in the absence of the setting itself. The simple knowledge that a place where one can enjoy nature is nearby may be a source of pleasure, perhaps explaining why the actual use of a nature setting is not essential in people's expressions of satisfaction. Such a *conceptual* form of involvement was one of the most important sources of satisfaction that participants expressed for a nearby vest-pocket park (R. Kaplan, 1980). The joys of gardening, in the dead of winter, are also necessarily more conceptual than "real." But planning for the future can seem real enough. Certainly, when residents vehemently rebel against impending development of the countryside their actions are based on the conceptual significance of such areas.

THE IMPORTANCE OF NEARBY NATURE

People are often aware that nature is important to them. They appreciate having daily contact with it, even if there is only a modest amount of "nature" nearby (Talbot & Kaplan, 1984*). When asked why natural areas are important to them, residents often indicate that they "enjoy them" or that they appreciate the beauty of nature. This was the most frequent response in the Talbot et al. (1987*) study. Other reasons included values that are closely tied to activities (place for kids to play) and the opportunity for gardening.

Several other reasons for appreciating nature areas reflected the variety of functions that such settings can serve. For example, with respect to their own front and back yards, the areas helped mark a territory; the respondents indicated that these were areas where they could do what they wished; they were their own "little spots." Other people's yards, by contrast, were seen as sources of community pride. For some of the slightly more distant nature areas, and ones they did not frequent often, the value was expressed in more conceptual terms. Knowing that such open areas are available, that they are not built up and should be preserved that way, was seen as a strong basis for their importance.

In this study as in some others as well (Bardwell, 1985*; Talbot & Kaplan, 1984*), participants talked about natural areas as places to think and to forget their worries, to regain sanity and serenity, and to enjoy solitude.

For many people, however, the satisfactions they derive from nature are not self-evident. Some study participants were apologetic when asked during an interview about their contact with nature and what they derive from it. It is often difficult enough to have individuals realize the extent of their interaction with things natural. To have them consider the many ways in which such interactions affect them can be an unreasonable task. Advertisers, for example, use imagery of natural settings extensively. They do not try to call attention to this device but no doubt realize the benefits of such a backdrop.

Research on the benefits of natural settings thus poses many dilemmas. Just as was true with the exploration of perception (chapter 1), we are dealing with a domain that is difficult to study directly. As was true before also, the results of any single study are easily challenged. The cumulative insights gained from a series of studies, however, reduce such skepticism. This is all the more the case given a series of studies that are carried out to examine different questions, in varied settings, and with participants who differ demographically. What such studies have shown over and over again is that nature is, in fact, important to people. Exploring why and how it is important has been an enlightening quest.

Residential Satisfaction and General Well-being

One indication of the importance people place on the availability of natural settings is that they are willing to pay for it. Rental rates are often higher for dwelling units that face a river or woods. Property values are supposedly increased by adjacency to golf courses. Even when such amenities are not present, the names of housing developments often reflect natural elements (e.g., Forest Hills, Meadow Creek, Woodbury Gardens, Pine Valley). We have carried out studies that have examined this relationship more directly by asking residents to answer some questions about their satisfactions.

The study that we discussed in the context of perception and preference research (in chapters 1 and 2), because it included a photo-questionnaire, also addressed a variety of issues related to the availability, adequacy, and importance of the nearby-natural setting (R. Kaplan, 1983, 1985a*). Nine multiple-family housing projects were included so that we could study differing kinds of nature settings and communities that varied in composition and size. Among the many questions included in the survey were items regarding neighborhood satisfaction. Using Category-Identifying Methodology (CIM), these items formed the basis for four "satisfaction" categories: Community, Size (of the development), Physical Facilities, and Nature.

The satisfaction with Community was particularly affected by the view of gardens and by the adequacy of having places to grow things. The other three satisfaction scales were all strongly affected by views of woods and trees as well as by the number of trees near the residence. Residents' perceptions of the adequacy of nearby-natural areas and of places for taking walks were strong predictors of each of these three satisfaction measures. These results indicate the broad impacts of nature availability as factors in satisfaction with both the physical and the social components of the residential setting.

Unlike the multiple-family housing study, Frey (1981*) included a wide variety of residential settings, drawn from 38 urban neighborhoods within the same city. Her findings are similar to the former study, however. Based on CIM procedures, her study entailed six aspects of neighborhood satisfaction, and four of these included items reflecting nature content. Nature thus not only emerges as an aspect of the physical environment but is identified with a wide range of attachments and potential involvements with the near-home environment.

In Frey's study, as with the multiple-family housing study, neighborhood satisfactions were strongly affected by aspects of the natural environment. She reported higher neighborhood satisfactions among respondents living in areas with more trees and certain other natural features, and among those who felt that their neighborhoods included specific natural features that they would miss if they moved elsewhere. Furthermore, physical proximity to such natural elements was positively related to neighborhood satisfaction. Thus aspects of both nature involvement and access to nature were important predictors of residential satisfaction.

Frey's study provided further indications that the effects of nearby nature extend beyond people's responses to the physical setting. Included in her study were several questions relating to life satisfaction. Each of the six categories of neighborhood satisfaction was significantly related to an individual's perceived degree of life satisfaction. In other words, higher levels of both neighborhood and life satisfactions were found among individuals who more frequently pursued gardening and other nature-related activities near their homes.

Fried (1982, 1984) reports findings that further corroborate this pattern of results. His study involved a carefully drawn, stratified, multistage area probability sample from 42 municipalities across the United States. He also found that nature items emerged as part of community satisfaction categories (using CIM procedures). The strongest predictor of Local Residential Satisfaction was the ease of access to nature (based on both ease of access to the outdoors in the neighborhood and the closeness of larger open spaces). The category dealing with Local Convenience Satisfaction, by contrast, included satisfaction with the availability of local parks and recreation.

Fried's findings are most revealing. He indicates that "the single dominant finding is that, even by comparison with such major variables as marital and work satisfaction, community satisfactions make a notable contribution to life satisfaction." The Local Residential Satisfaction grouping was second most important, after Marital Role Satisfaction, in predicting life satisfaction. Even more striking are the results that show this pattern is most evident at the lowest status levels. Thus, for lower social class positions, the satisfaction with the physical setting is even more powerful in explaining life satisfaction than is the case as social status increases.

Relatively few studies have examined residential satisfaction in terms of physical, as opposed to social aspects. The studies discussed here are important in showing that different natural aspects of the immediate residential context play important roles in a variety of aspects of satisfaction. Furthermore, it is generally not the large open spaces and designated parks that contribute to satisfaction as much as such elements as trees, landscaping, and opportunities for gardening. In terms of Alexander's "3-minute rule," these results suggest that nature that is most immediately available does, in fact, make the most difference. Last, and most dramatic, are the findings that neighborhood satisfaction based on such physical aspects of the setting does, in fact, affect residents' general well-being.

Job Stress and Physical Health

Given the positive influence of natural elements in the residential context, it is reasonable to ask whether nearby nature can also increase satisfaction in the work setting. We are aware of very little research on this question. Certainly, some enlightened businesses have incorporated natural elements in and near the work environment. Perhaps the common use of big potted plants in lobbies also reflects at least an intuitive recognition of the role of natural elements.

"Stress management" has become an acknowledged necessity in the work setting, and corporations have expended vast sums on stress workshops for employees. An analysis of such procedures, however, shows a virtual neglect of the role of the natural environment (Tyll, 1988). In a few cases, stress management and relaxation procedures encourage individuals to imagine themselves in a natural setting. There are also sound tracks of nature sounds that are advertised as having soothing consequences. All these "uses" of nature, however, are remote. Whether nature at the doorstep of the workaday world can provide positive effects for employees remains a largely unanswered question.

We have begun to study this question (S. Kaplan, Talbot, & Kaplan, 1988). In our initial effort we included employees of a private, fast-growing corporation, as well as individuals who work for government agencies. None of these samples is very large, nor can we claim that those who responded are necessarily representative of their organizations. Since the study relied upon a self-administered questionnaire that inquired about job stresses and job satisfactions, great care was exercised to protect the identities of the respondents. Participants also answered questions regarding their view from their desks (if their job involved one) and other aspects of their setting.

The results of the Job Pressures Project have been remarkably consistent with our expectations. Access to nature at the workplace is, in fact, related to lowered levels of perceived job stress and higher levels of job satisfactions. The findings indicated that workers with a view of natural elements, such as trees and flowers, felt that their jobs were less stressful and were more satisfied with their jobs than others who had no outside view or who could see only built elements from their window. Furthermore, employees with nature views also reported fewer ailments and headaches.

The results of the various studies provide strong support that nearby nature affords a wide range of both psychological and physical benefits. People feel more satisfied with their homes, with their jobs, and with their lives when they have sufficient access to nature in the urban environment. People value natural settings for the diverse opportunities they provide – to walk, to see, to think. They are not necessarily aware of the many forms of encounter they have with nature or of the variety of benefits that accrue.

THE MANY KINDS OF NATURE

These satisfactions derive from contact with many forms of nearby nature. The notion of *open space* is often used interchangeably with *natural areas*. The series of studies that are the basis of much of this chapter, however, lead us to suspect that these are quite different. Open spaces often serve as places for ball playing and Frisbee, for flying kites, for larger social gatherings. Though such areas are useful, they often show much lower preferences than areas that are less open (Bardwell, 1985*; R. Kaplan, 1985a*; Talbot et al., 1987*). Large open areas in residential settings can also be problematic because they are more difficult to monitor. Newman (1972) coined the expression *defensible space* to refer to areas that are under community control and surveillance. This is more easily achieved if the open spaces are not too large and clearly "belong" to a particular cluster of residences.

On the other hand, natural areas such as fields or woods or a pond or marsh that are not too far away often play a special role for people. Even if they are not used frequently, their potential availability makes an important contribution. Such areas are often appreciated for their "thereness." The knowledge that one could enjoy such an area is in itself a source of satisfaction. This issue is particularly important when satisfaction is assessed in terms of usage (Francis, 1987b). By counting only the number of visitors or interviewing only those in a park, one can easily lose sight of the fact that for many individuals the knowledge of the park's existence is a source of pride and satisfaction (R. Kaplan, 1980; Ulrich & Addoms, 1981).

Nearby nature, as we have seen, need not be a big area. In fact, even the sight of a few trees, the view from a window at home or at work, can provide satisfaction. Getz et al. (1982) used a variety of approaches to assess the preferences of inner-city residents for trees, woodland, and open areas in the urban context. The scene of the treeless business district fared least well, but a tree standing alone received more positive ratings than either the woodland or the tree-lined residential street scenes. Stribley (1976) found residents at a public housing project were particularly dismayed by the lack of trees in their immediate setting. Not only are the trees themselves engrossing, but the birds and squirrels that use them also add to the delight.

Similarly, landscaped areas can also be a source of satisfaction, whether or not one participates in their maintenance. Not only are the individual species often interesting, but the landscaped area creates a "space," a setting that can be a source of satisfaction to be in or to see.

Another important kind of natural area is the "special spots" (Frey, 1981*) that one feels very possessive about and considers to be one's very own. Often memories of favorite childhood places, such as a grandmother's garden (Francis, 1987a), include some special place – perhaps where one could hide from the adult world. In one's current surroundings, a particular tree may be the source of great individual affection, possibly because it has a special form or affords a comforting view or provides a comfortable place to sit and pause. How rarely any of us know of others' special places. In fact, one often does not become aware of such attachments until a threat arises.[2]

There is yet another kind of natural area that is nearby and highly valued: the garden. Gardens here refer not to magnificent places that are professionally maintained but to plots of land where individuals grow plants of their choosing. Other people's gardens can be the source of great pleasure, and, in fact, the views from many residences are to neighboring gardens. As already mentioned, the satisfaction with Community was closely related to the view of gardens, one's own and others'.

Gardens can come in many sizes and can be grown in many places - even on rooftops. There is probably no single nature-based activity that is so widely shared by the population. People who garden come in every color, size, shape, nationality, and income level. People garden whether they live in rural areas, in the suburbs, or in the innermost, built-up, teeming portions of cities. They do it individually, in family groups, or as part of a community.

In fact, gardening is an amazing phenomenon. Why should this activity be so popular? Certainly, the opportunity to grow fresh vegetables and fruits is an attraction. But many gardeners do not grow edibles. Furthermore, there is no guarantee that one's efforts will bear fruit. Marauders, human or otherwise, are no trivial threat. The weather is a further unpredictable element with threat potential. There are plenty of other negatives to this activity: It is hard work and gets one dirty, sweaty, and achy. It is potentially expensive. It is often hard to know how much to plant. The garden needs so much attention and often looks messy nonetheless.

Despite all this, gardens spring forth everywhere, and the legions of gardeners grow as do their plots. It is worth examining this kind of contact with nature more closely.

THE SPECIAL CASE OF GARDENING

The "double standard of open space" of which Little (1974) wrote referred to the inequitable distribution of "greenery" in the United States. "The logic of our policy seems to rest on this syllogism: inner cities have no greenery; poor people live in inner cities; therefore parks, open space, and wilderness are not necessary for them." Indeed, he provided a perceptive analysis for the woeful absence of nature in the urban, especially inner-city, context.

That picture has changed dramatically. In numerous cities, and especially in inner-city areas, greenery is now part of the neighborhood. The story of community open spaces and community gardens is being told across the country. Unlike so many other environmental topics, this one is frequently a success story. Newspapers often carry articles of local, grass-roots projects that yield tomatoes and neighborhood transformations in areas where many had feared to tread.

The National Gardening Association (1985) reported on three separately conducted surveys of community gardening. These provide estimates of the extent of this effort, in this country, during 1984–85: over one million households, 91,000 acres of land, 12,000 garden sites, and more than 10 times as many individual plots. Virtually every state has examples of community gardens, in many instances supported by municipal agencies. An effort of this magnitude requires many forms of support. In some cities, handbooks have been developed to help new groups get started (e.g., Naimark, 1982). Francis, Cashdan, and Paxson (1984) supply a useful summary of the pervasiveness of the movement both in Europe and in this country.

Fox, Koeppel, and Kellam (1985) provide a bibliography of over 50 organizational resources useful in land acquisition, grass-roots fund raising, as well as the horticultural aspects of such projects. The American Community Gardening Association includes as its members organizations such as Operation Greenthumb (New York City), SLUG (San Francisco League of Urban Gardeners), Common Ground (Los Angeles), BUG (Boston Urban Gardens), and CitiParks (Pittsburgh). The National Community Garden Preservation Program (sponsored by the Trust for Public Land) provides technical assistance to local groups who are trying to save community gardens from sale or development. Other groups offer grants and technical assistance with respect to other facets of these projects.

Well before the current interest in community gardens, the New York City Housing Authority sponsored its annual Tenant Gardening Competition. Even in the mid-1970s, this activity involved over 14,000 people of all ages. For many years, the competition included only flowers, but even when vegetables were permitted flower gardens were still the more prevalent. The tenants in these public housing projects tend to be people for whom many things are not going well, for whom despair is a way of life. If one subscribes to some hierarchical notion of human needs (e.g., Maslow, 1962), one would hardly expect such people to invest themselves in raising flowers when the more basic requirements of living are in such shambles.

As early as 1972 and in a series of later publications,

Lewis documented the pervasive effects of these gardening projects. As a judge for the competitions, he recognized that much more was happening here than the growth of pretty blossoms. From the perspective of "beautification," of making these rather desolate areas look more attractive, the effects of the competitions can be viewed in terms of "amenities." But Lewis's perceptive analysis left no doubt that these flower patches brought with them many other benefits as well.

To guard the cherished plants, neighbors organized window watches with scheduled shifts so that upper-floor residents could mobilize ground-floor "co-workers" in the event that protective action was needed. Thus the gardening projects required cooperation among tenants and even some talent in organizing. Lewis also tells of elaborate schemes where adjacent areas were painted to coordinate with the flowers. Thus the area that was transformed was much greater than the flower garden itself. The results of these various efforts were indeed cleaner areas - one would not want a littered sidewalk in front of prized zinnias - and generally more attractive housing projects. In the public housing context these "improvements" reflect community pride, reduced stigma, and substantially enhanced self-esteem.[3]

THE PSYCHOLOGICAL BENEFITS OF GARDENING

From Lewis's sensitive observations it is clear that the tenant flower competitions nurtured more than flowers. The mere sight of these areas provided pride and joy, even to those who did not participate directly. For the active participants, the benefits were even more extensive. From observations and casual exchanges with these and other gardeners one can learn a great deal. The conference proceedings on the "meanings of the garden" (Francis & Hester, 1987), for example, provides a rich assortment of insights about the many ways in which gardening makes a difference to people.

Ideally, however, one would also have some more systematically collected data to document the types of benefits derived from this activity. Unfortunately, there has been relatively little research on the psychological dimensions of gardening. Here again, as with any research, the choice of method can make a difference. For example,

Francis (1987a) has collected information on the meaning of gardens by interviewing gardeners both in California and in Norway. From this information he extracts some themes about why gardens play such an important role. One of his conclusions is that "much of the meaning of garden for people can be traced to the concept of control. The garden is a place that people can directly shape and control in a world and environment largely outside their direct control."

Our own research on the benefits of gardening also included the question of "control." The approach we took was more structured, including a series of items about potential source of satisfaction and requiring ratings on a 5-point scale. Such an approach has the disadvantage that one is "putting words in the mouths" of the participants – in other words, they are not generating the sources of satisfaction themselves. On the other hand, comparability across respondents is enhanced if everyone is addressing the same questions. With respect to control, for example, we found that it was a meaningful concept but of far less significance to our participants than many other aspects of gardening benefits.

Such control is, of course, illusory. Though the gardener can exert some control, despite the best intentions plants do not always heed such instructions. In fact, experienced gardeners often find the plants do some of the controlling. The microclimates of a garden can lead to far better results in one spot than in another, even if one's plans were otherwise. If one's vigilance is not constant, one might find plants have migrated to places of their own choosing. Participation may be a far more appropriate concept here than control.

Before discussing some of the benefits of gardening that our research showed to be of greatest importance, we should provide a brief description of the studies that addressed these questions. The initial effort, in the early 1970s, was on a small scale, when community gardening was just beginning to take hold.[4] The study included individuals involved in community gardens as well as backyard gardeners (R. Kaplan, 1973a). The participants were not representative of their respective category, nor was the study carried out on a large scale. The findings, however, were most useful in identifying some facets of gardening that had received no prior research attention.

A much larger sample was involved in a later study (R.

Kaplan, 1983; R. Kaplan & Kaplan, 1987). This study was carried out in collaboration with Charles Lewis and the People/Plant project, which the American Horticultural Society (AHS) then sponsored. A two-page questionnaire was sent to the AHS membership in 1976; the 4,297 responses came as an overwhelming surprise. If a structured questionnaire is restricting because it only requires rating scales, this was not evident from the returns. Many appended letters and comments. The significance of gardening was expressed in many ways, some eloquent, some simple, some with sadness in anticipation of moving, and others with a clear mission (e.g., "Grow garlic!"). In any event, there was no way to respond to this many letters on an individual basis. Perhaps even more frustrating was the fact that the results of the study were never made available to the participants because AHS decided not to publish the study's results in the magazine that reaches its membership.

The AHS members tend to be more affluent than the average gardener and more inclined to grow flowers (even exotic species) than vegetables. For these reasons, we included a very different group of gardeners for comparison purposes. With the help of Jerome Goldstein, then with *Organic Gardening and Farming* (OGF), the magazine's readers were invited to join the project. A total of 240 readers responded and returned surveys. The OGF sample tended to be younger, distinctly less affluent, and more oriented to growing vegetables than the AHS participants. Both samples, however, included individuals from all regions of the country (all states in the case of AHS). Once again, it should be emphasized that there is no basis for claiming representativeness here.

In both the earlier study and the AHS/OGF project, a clear area of satisfaction derived from gardening involves the Tangible Benefits. The items in this category (using the same Category-Identifying Methodology discussed in previous chapters) were the same (importance of growing one's own food, cutting expenses, and the satisfaction of harvest). As would be expected, the OGF members considered this a particularly important facet of satisfaction. In each of the studies, however, the importance of these tangible aspects, as well as the emphasis on vegetables, changed with experience. As gardeners gain more experience, the interest shifts to flowers, and the Tangible Benefits category becomes less salient.

Table 5.1 shows the mean ratings of each of the CIM-derived themes for the AHS and OGF groups. (The specific items are also indicated for each of these themes.) Whether the person was strongly oriented to growing the family's food or to raising prizewinning exotic species, the greatest benefit of gardening was related to the sense of tranquility it afforded. As a matter of fact, the satisfaction derived from Peacefulness and Quiet was, from a statistical perspective, significantly greater for the OGF than for the AHS sample. (The initial study did not include items related to this theme.)

Table 5.1. *Mean ratings for gardening satisfaction categories*

	AHS	OGF
Peacefulness and Quiet (feeling of peacefulness; source of quiet and tranquility)	4.3	4.5
Nature Fascination (likes the planning involved; gets completely wrapped up in it; never fails to hold interest; seeing plants grow; checking to see how plants are doing; working in the soil; working close to nature)	4.2	4.4
Tangible Benefits (producing own food; cutting food expenses; harvesting)	3.4	4.3
Sensory (walking in the garden; creating something beautiful; colors and smells)	4.3	4.2
Share-Tangible (sharing produce or flowers; being able to give others things I grew myself)	3.7	3.8
Novelty (trying new kinds of plants; growing odd and unusual plants)	4.1	3.7
In My Control (the one place where it's up to me how it looks; something I can do on my own; proving to myself what I can do)	3.5	3.6
Share-Knowledge (having people come to me with gardening problems; sharing gardening information; helping others get started in gardening)	3.4	3.4
Tidy and Neat (how neat and orderly things look; fixing up that it leads to; keeping everything tidy and nice)	3.2	3.0

Although gardening is often justified in terms of the joys of the harvest, both practical and experiential, somewhere along the way other forces seem to become even more important. The initial study led us to see that fascination plays a central role among the sources of gardening satisfaction. For the subsequent, larger study, therefore, we incorporated substantial additional material related to the potential importance of fascination. As is clear in Table 5.1, Nature Fascination was one of the strongest sources of satisfaction for each sample.

The items in the Nature Fascination category are interesting for their diversity and for their relationship to nature. Some of the items reflect a physically active involvement (such as working in the soil) whereas others focus more on observation (checking on the plants) and still others are at a more cognitive level. An important source of satisfaction derived from gardening is thus that it holds one's attention in a multitude of ways, even when the garden lies dormant.

Fascination receives little recognition as a psychological benefit, even though the research findings repeatedly point to its importance. As was true with the perception and preference research discussed in part I, here again the interplay between theory and empirical study has been exciting. The theoretical perspective made it easier to recognize the significance of fascination; the data made it easier to understand its multifaceted expression. S. Kaplan (1977, 1978) wrote of the importance of fascination in achieving cognitive clarity. In chapter 6 we will return to these ideas.

One other aspect of the results of the larger gardening study is particularly intriguing and relates back to Francis's (1987a) comments about the importance of the sense of control that gardening affords. Among the AHS respondents, 27% indicated that they use mostly chemical fertilizers, 30% used mostly organic, and 41% replied that they use "about the same" of each. (For the OGF sample, only 3% indicated "mostly chemical.") An important reason to favor chemical fertilizers is as a form of control. Organic gardening tends to be much lower in the instant cures promised by the advertisers of garden chemicals and would thus involve a reduced degree of control. In the absence of fast cures, however, the nonchemical gardeners may be more vigilant, more observant of potential problems. In fact, for both AHS and OGF groups there was a tendency for those who used organic fertilizer to spend more time in gardening, though there were no differences in size of gardens.

The differences among the AHS gardeners were remarkably (and significantly) large, based on their indicated fertilizer preferences. Those who tended to use chemical fertilizers indicated consistently lower satisfaction ratings on each of the benefit categories. In addition, their rating of a pair of life-satisfaction items (e.g., "How satisfying would you say your life has been compared to other people you know?") was also significantly lower than for the "organic fertilizer" and "both chemical and organic" groups. The latter two groups were similar in life-satisfaction ratings and on several of the gardening satisfaction themes. Those who used a mix of fertilizers scored higher on satisfaction from novel plants, sharing information, and having a tidy and neat garden. The organic fertilizer group, by contrast, rated higher with respect to Tangible Benefits as well as Peacefulness and Quiet.[5]

Gardens provide a clear example that the nearby-natural setting does not need to have great extent. Even very small gardens provide many of the benefits discussed here. Gardens also show the diverse "uses" of the nearby-natural setting, including active, physical effort as well as less direct involvement. Furthermore, in the case of gardening, as opposed to other outdoor activities, many participants are aware of the benefits that such efforts provide. The awareness, however, even in the case of gardening, is restricted to certain domains. The joys of the sights and smells, and even the peacefulness, may be more apparent than the benefits that accrue from the fact that gardening provides an activity that is so characteristically engrossing and captivating.

CONCLUDING COMMENTS

It may be useful to reexamine the description of the urban end of the nature continuum in the Recreation Opportunity Spectrum (ROS) and to explore a modified description, which the findings presented in this chapter would suggest (Table 5.2). Both with respect to the setting and with respect to the experience we would suggest that nearby nature provides a much wider range of opportunities. In

Table 5.2. *A contrasting view of the ROS: urban context*

ROS	ROS revisited
Setting characterization	
Area is characterized by a substantially urbanized environment, although the background may have natural-appearing elements. Renewable resource modification and utilization practices are to enhance specific recreation activities. Vegetative cover is often exotic and manicured. Sights and sounds of humans, on site, are predominant. Large numbers of users can be expected, both on site and in nearby areas. Facilities for highly intensified motor use and parking are available with forms of mass transit often available to carry people throughout the site.	Area is characterized by vegetation in context of built environment. Lawns are frequent form of vegetation. So are trees and shrubs; these may be sparse or continuous (street trees) or in greater abundance (parks and woodlots). Vegetation may be carefully arranged and maintained, as in a landscaped area or garden, or less intentional, as in an unused lot. Use levels vary from none or few to heavy usage. Much of nearby nature entails no specific facilities. Access may be by car, bike, or on foot; some areas are private (back yards).
Experience characterization	
Probability for experiencing affiliation with individuals and groups is prevalent, as is the convenience of sites and opportunities. Experiencing natural environments, having challenges and risks afforded by the natural environment, and the use of outdoor skills are relatively unimportant. Opportunities for competitive and spectator sports and for passive uses of highly human-influenced parks and open spaces are common.	Much nearby nature is experienced with only one or two others or by oneself. In fact, opportunities are often sought to get away from others. Risk and challenge are not as critical as are opportunities for observing and a context for thinking. The experience of nature itself is often an essential aspect. Many such settings are seen as restorative in their lack of bustle and as sources of multiple forms of fascination.

many ways, in fact, it is hard to distinguish these benefits from those obtained in the more remote and primitive settings.

Urbanization and population growth have created many stresses. These trends have created new pressures and made some of the old satisfactions harder to achieve. Peace and quiet, fascination, the chance to share with others and to do what one wishes are all deeply important to human beings. The natural setting makes these satisfactions more available. Even the view of trees can lead to psychological gains. Understanding that the most readily available patch of nature can serve to facilitate such gains is of utmost importance. Although we do not deny that urban parks and ballfields are also useful, they simply cannot and must not be seen as serving many of the urgent needs of urban people.

The immediate outcomes of contacts with nearby nature include enjoyment, relaxation, and lowered stress levels. In addition, the research results indicate that physical well-being is affected by such contacts. People with access to nearby-natural settings have been found to be healthier than other individuals. The longer-term, indirect impacts also include increased levels of satisfaction with one's home, one's job, and with life in general. Surely this is a remarkable range of benefits for such a relatively simple and inexpensive environmental change.

NOTES

1. Fortunately, there is a growing literature on nature in the urban context. A substantial amount of research in this area has been supported through the Urban Forestry Project of the U.S. Forest Service; Schoeder (1988) and Smardon (1988) have written useful reviews of some of these studies. A special issue of *Landscape and Urban Planning,* edited by Rowntree (1988), concerns "urban forest ecology." It includes several papers on the values people place on urban vegetation. Related research on urban nature and the role of nearby open spaces has also been reviewed recently by Francis (1987b) and Hayward (1989).
2. Threats can come in many forms. The Dutch elm disease epidemic left many people quietly mourning the loss of a favorite tree. There is even a tree whose demise was recognized by the entire community. When a huge white oak (120 feet tall and 237 inches in diameter) fell over in a storm in 1987, citizens of Copper Harbor, Michigan, declared a day of mourning, "complete with memorial service." The tree's age was estimated to be over 600 years, predating even the struggles of the early settlers. One member of the Board of Commissioners commented on the importance of the day of mourning because "not only was it a beautiful tree, but when you saw it, you could feel the peace and history" (National Urban Forest Forum, 1988).

3. Similar benefits have been reported in the context of prisons where inmates have had the opportunity to work on farms or gardens. In fact, there is a publication, *The prison garden book* (Flinn, 1985), which is intended for prison adminstrators and inmates. Such gardens have been found to be cost-effective from an economic point of view. They have also played an important role in job training. Perhaps their most important consequence, however, is in terms of changes in even the most hardened of prisoners. They find the work meaningful and are willing to assume responsibility; they experience accomplishment rather than failure, and self-esteem is notably improved.
4. The recent resurgence of community gardens can hardly be viewed as a "new invention." A similar tradition has been strong in Europe for a long time, and the Victory gardens of the 1940s were also similar in many respects.
5. The organization that sent the questionnaire to its membership later reneged on its promise to publish the results of the survey on the grounds that they would simply be of no interest to its members. One cannot help but wonder whether the difficulty stemmed not so much from there being nothing illuminating in these results as from the content of these findings (especially those related to chemical fertilizer use).

PART III

TOWARD A SYNTHESIS

In the loveliest town of all, where the houses were white and high and the elm trees were green and higher than the houses, where the front yards were wide and pleasant and the back yards were bushy and worth finding out about, where the streets sloped down to the stream and the stream flowed quietly under the bridge, where the lawns ended in orchards and the orchards ended in fields and the fields ended in pastures and the pastures climbed the hill and disappeared over the top toward the wonderful wide sky, in this the loveliest of all towns Stuart stopped to get a drink of sarsaparilla. (E.B. White, *Stuart Little*)

THE imagery of nature provides a sense of wholesomeness, of repose. Nature can be inspiring, awesome, tranquil, or calming. One can be absorbed by the blossoms of an African violet on the windowsill or cascading waterfalls or the rustle of wind in the trees. People in all walks of life, in sickness and in health, in good times and bad times, find in nature something that comforts and restores.

Part III explores the concept of a restorative environment. Drawing on the research on environmental preference and on the satisfactions and benefits of nature in the nearby as well as more remote settings, it provides a way of understanding the human–nature relationship. An examination of the factors that make a restorative experience more likely provides the basis for integrating the several themes of this book.

Part III goes beyond the fact of the restorative power of nature – near and far – to examine the means whereby this effect is achieved and the properties of the environment that make it possible. Such information turns out to be of great interest to those many individuals – the gardeners, the hikers, the bird watchers, the campers, and the many other outdoor people – who treasure such experiences and have wondered about the source of these profound effects.

The information may also serve some more practical functions as well. Three of these come immediately to mind:

1. There is justification here for those concerned with conservation of natural environments, big and small, from the vacant city lot to the scenic landscape.
2. Those in the mental-health-related professions can add a further dimension to their repertoire of mind-healing procedures.
3. There is potential insight here for the designer, the planner, the manager. There are innumerable opportunities for creating, enhancing, and preserving the restorative potential of natural settings. Some progress of this kind has, of course, already been made. There is reason to think, however, that a conceptual framework may make such efforts more effective and more consistently practiced.

It is naive to assume that an emphasis on restorative experience and on the restorative environment that supports it will be greeted enthusiastically by all concerned. There will be (and are) those who will argue that people are doing fine without all this and that such an approach is merely a flimsy excuse to hinder material progress and economic development. The blunt fact of the matter, however, is that people are not doing fine. Difficulties abound, even in an affluent and enlightened society like ours. Current levels of family trouble, child abuse, and homicides are painful clues that much is not as it should be.

There are undoubtedly many factors underlying this troubling state of affairs. We have argued elsewhere that confusion and frustration in many facets of life have contributed to stress (S. Kaplan & Kaplan, 1978) and that people would be more likely to thrive if the environment were more supportive of their goals and inclinations (S. Kaplan, 1983). In the long run – if there is a long run – these problems need to be addressed directly. But, for the foreseeable future, restorative environments offer a concrete and available means of reducing suffering and enhancing effectiveness.

CHAPTER 6

THE RESTORATIVE ENVIRONMENT

WHEN pressures have built to a critical point people say they have "to get away from it all" or "to escape." These expressions suggest the need for a change of venue, but they ignore the fact that where one is headed may be as important as where one is coming from. One can, after all, escape to many places that would fail to achieve the desired recovery.

Being away may well be a requisite condition for a restorative experience, but it does not address other necessary features. The purpose of this chapter is to examine the qualities that characterize a restorative environment and to explore the psychological benefits that such environments make possible.

The question of what makes an environment serve a restorative function for a frazzled, hassled, worn-out, or irritable individual cannot be answered without first dealing with a prior question. It is necessary to address the issue of what constitutes the "mental fatigue" that a restorative environment is to help one recover from.

The wilderness research (discussed in chapter 4) played a particularly important role in the development of the ideas about what constitutes a restorative environment. In the context of that research we also began to examine the puzzles of mental fatigue more closely. As a framework emerged, it became apparent that the results of many of the other studies (particularly the gardening satisfaction research discussed in chapter 5) were equally applicable. Thus the synthesis that we offer in this chapter looks at the restorative environment in general, with a special focus on the natural environment. Since recovery from mental fatigue occurs at many levels in terms of both time and place, the restorative environment too can be small or vast, brief or more extended.

One more comment, by way of caution: In each of the previous chapters the discussion has been based more or less closely on research findings. The purpose of Part III is more conceptual. Based on the insights of the research on perception and preference as well as the benefits of natural settings near and far, the goal now is to find a way to understand not only that material but also situations that have not yet been studied. Thus this chapter delves into theory, into possible reasons (or mechanisms), into ideas as opposed to facts.

MENTAL FATIGUE

At the end of the work day or the work week or after months of steady effort one feels worn out, ready for a break or respite. In fact, all of us have experienced such a state, even in a matter of a few hours, such as after a protracted meeting, intense effort to complete a project, a period of worry and anxiety, or even just a period of trying to do too many things at the same time. The worn-out state in these situations is generally not physical; in fact, one might even complain of a lack of physical activity. Rather, these situations involve what we are calling *mental fatigue* (S. Kaplan, 1987b).

The concept of stress is used in many situations that we would describe as mental fatigue, but the two concepts are distinctly different. Stress involves the preparation for an anticipated event that has been evaluated as being threatening or harmful. Though mental fatigue may well result from such circumstances, it *also* arises out of hard work on a project one enjoys. In such cases there is no threat of harm, no negative evaluation, and no anticipation. As the personnel director of a local firm put it, "You don't have to have a bad boss to get worn out." Conversely, being chased by a bear is likely to be a most stressful experience. It need not, however, result in mental fatigue.

Calling this state mental fatigue, however, does not explain what is wearing down or becoming fatigued. What are the common consequences that result from long hours of study, too many late nights at the office, or days filled with worry and concern? Surely, to say that it is the mind that is fatigued is neither helpful nor even accurate. People so mentally fatigued that they can hardly function can nonetheless spring into action when an emergency arises or even when something of particular interest happens to come along. It would be a mistake to say that they were not genuinely fatigued simply because they can function well under certain circumstances. And it would be a mistake to attribute the fatigue to the mind as a whole when the individual is thoroughly competent in certain situations.

Two Kinds of Attention

A clue to this puzzle comes from the mental inertia, the difficulty of focusing, that is so characteristic of mental fatigue. For a mentally fatigued person, paying attention to something uninteresting is burdensome, but attending to something of great interest poses no particular challenge. This discrepancy parallels a seminal distinction first made by William James (1892). He identified two types of attention, distinguished in terms of the effort involved in their use. One type, which he called *involuntary attention,* refers to attention that requires no effort at all, such as when something exciting or interesting happens and we look to discover what is going on. By contrast, forcing oneself to pay attention to something that is not particularly interesting requires a good deal of effort. Though James called this kind of attention *voluntary,* this terminology has created a good deal of confusion; thus we shall adopt the expression *directed attention* instead.

James described stimuli that bring forth involuntary attention as having a "direct exciting quality." With characteristic exuberance, he listed examples of such stimuli: "strange things, moving things, wild animals, bright things, pretty things, metallic things, words, blows, blood, etc. etc. etc." Directed attention, by contrast, is not tied to particular stimulus patterns in the same way. One's efforts may involve many specific contents, but the attentional component is generic, or content-free.

It may be useful to explore a potential mechanism for this distinction to help conceptualize the mental fatigue process. A key issue here involves the concept of *inhibition.* As James envisioned it, the way one maintains one's focus on a particular thought is not by strengthening that particular mental activity but by inhibiting everything else. From this perspective, the greatest threat to a given focus of attention is competition from other stimuli or ideas that can be the basis of a different focus. Inhibiting all such potential distractions protects and hence sustains the original focus.[1]

In focusing, or directing, attention, a great deal of effort is devoted to avoiding distractions or as James put it, to "resisting the attractions of more potent stimuli." Thus, although one's efforts may involve very different tasks, the inhibition of the ever-present distractions is generic. Rather than calling upon a different pattern of inhibition for each different thing one pays attention to, the same directed attention mechanism is called upon again and again. Given the frequency with which such an inhibitory process must be called upon, it is easy to see that directed attention would be susceptible to fatigue. Simply put, then, it is our hypothesis that when one experiences mental fatigue the underlying cause is fatigue of directed attention.[2]

From this perspective, the reason that emergencies and other interesting events can rouse the mentally fatigued individual is that these circumstances are high in involuntary interest and hence do not require directed attention. In fact, since the involuntary attention can sustain itself, the fatigued directed attention has an opportunity to recover.

The Costs of Directed Attention Fatigue

At first glance a decline in capacity for directed attention might seem unfortunate but hardly catastrophic. The role of directed attention in human functioning is, however, far more pervasive and more crucial than it might seem. Consider what would happen if one were unable to exert inhibitory control in perception and action. It would be impossible to focus in the face of distraction; thus many tasks could be carried out only under the most ideal of circumstances. Some tasks, where distractions are intrinsic (such as air traffic control or proofreading or even driving in the city), could not be achieved at all. It would be impossible to pursue purposes, since this requires keeping a goal in mind despite the challenge offered by an often uncooperative environment.

In addition to influencing what we pay attention to, this inhibitory capacity also underlies our ability *not* to respond. There are times when it is advantageous to delay, to inhibit action until the appropriate time or until sufficient information is available. Not responding also gives one time to think. Being unable to resist responding thus precludes taking advantage of the power of our cognitive apparatus. Being unable to inhibit one's response also inclines one to act in an impatient and risky fashion. A mentally fatigued individual is far more likely to commit what accident investigators call a "human error."

There is another major category of behavior that suffers seriously from the fatigue of inhibitory control. Socially responsible behavior often requires inhibiting personal feelings and concerns in favor of group norms and values. An individual unwilling or unable to do this is unlikely to be an acceptable member of a group. Such a person, lacking the support and protection provided by group membership, would be in a most vulnerable situation should some threat or hardship arise.

Thus, without directed attention one is likely to be rash, uncooperative, and far less competent. Early hominids who *enjoyed* being in a confused state, nonchalant about the demands of focusing attention, would have been far less likely to survive and hence would have been unlikely to have become our ancestors. Those who found this state of mind unpleasurable and thus presumably were motivated to avoid it would have had an adaptive advantage. Extreme levels of mental fatigue, therefore, not only lead to incompetence but are painful as well.

Irritability is one likely expression of such a fatigued state. It expresses itself all too often and dangerously in crowded urban environments. But it is by no means the whole story of damage to social relationships caused by mental fatigue. A number of ingenious studies suggest just how unpleasant a mentally fatigued person can be. Those studies have in common that they require participants to carry out attention-demanding tasks under conditions of high distraction. After exposure to such an experience individuals are less likely to help someone in need (Cohen & Spacapan, 1978; Sherrod & Downs, 1974). They are also more aggressive (Donnerstein & Wilson, 1976), less tolerant (Cohen and Spacapan, 1978), and less sensitive to socially important cues (Cohen & Lezak, 1977). Thus the consequences of fatigued attentional capacity are strikingly pervasive. It is also interesting that, although the fatigue in these experiments was induced by a task that made heavy demands on attention, many of the effects are not specifically attentional, but are, as we might expect, expressions of a rather generalized decline in inhibitory control.[3]

Whereas directed attention must have played an important role for our ancestors, there are reasons to believe that it has become even more essential for functioning in the modern world. Increasing specialization has meant that each of us spends longer hours pursuing a single activity, as opposed to the variety of tasks pursued by our ancestors. Such persistence requires discipline, which depends heavily on directed attention. But not only have roles changed; the environment has changed as well. To the extent that what is interesting in the environment is what needs to be attended, one can function without the aid of directed attention. Though this perhaps was a common occurrence for early humans, in modern times many factors have contributed to a considerable gap between what is interesting and what is important. Population growth, urbanization, and industrialization have created a world where much of what is essential to pay attention to is far less interesting than are the competing stimuli.

THE RESTORATIVE EXPERIENCE: SOME KEY COMPONENTS[4]

The struggle to pay attention in cluttered and confusing environments turns out to be central to what is experienced as mental fatigue. If mental fatigue is the result of an overworked capacity for directed attention, then resting this capacity would seem to be the route to recovery. One way to achieve this is through sleep, and sleep is indeed a popular activity among fatigued individuals. Sleep, however, has limitations as a way to achieve recovery. Ideally, one would provide rest for directed attention during one's waking hours as well. Achieving this requires environments and tasks that make minimal demands on directed attention.

There are, no doubt, many kinds of environments that can be restorative. The wilderness research discussed in chapter 4 played a central role in our conception of key components that contribute to the restorativeness of an environment. The effects of the wilderness environment were powerful, and the experience was deeply restorative.[5] Reflecting on these findings and attempting to relate them to research in outdoor recreation lead to the identification of four central aspects of restorative settings.

Being Away

There is a tendency in the recreational literature to equate the idea of a restorative experience with escape or withdrawal (Driver, 1972; Driver & Knopf, 1976; Hollender, 1977; Ittelson, Proshansky, Rivlin, & Winkel, 1974; Klausner, 1971; Stringer, 1975). This interpretation has a certain intuitive appeal. People seeking a restorative experience speak of needing to "get away" and may describe the desired experience as a "change and a rest." As a theoretical or explanatory concept, however, it leaves much to be desired.

The term *escape* usually refers to an absence of some aspect of life that is ordinarily present, and presumably not always preferred. One might, for example, escape from crowds, noise, or routine. From an informational point of view, there are at least three different patterns that fit this description. A person might, for example, get away from distraction. Although this may bring to mind a retreat on a faraway hillside, in the literal sense of *escape* a quiet basement lacking a telephone could serve just as well. Another form of getting away involves putting aside the work one ordinarily does. Here the escape is from a particular content, and perhaps from anything that might serve as a reminder of that content. A third kind of escape is more internal in origin. It involves taking a rest from pursuing certain purposes, and possibly from mental effort of any kind. A given instance of escape might, one would suppose, involve one or more of these aspects. Presumably, the strongest effect would be achieved by combining all three of them. Yet it is difficult to believe that even such a combination would necessarily have a restorative impact. Certainly there are no lack of environments in which distraction is minimal, familiar contents are absent, and one's customary purposes cannot be pursued. Such settings might nonetheless be confining or boring or both.

Other Worlds and the Concept of Extent

To escape to a phone booth or a prison cell would hardly be restorative. Such settings may provide change, but they are greatly limited in scope. By contrast, restorative settings are often described as being "in a whole other world." Two properties are important to this experience: connectedness and scope, together defining the concept of *extent*.

The sense of being in a whole other world can arise in many contexts. Playwrights use it extensively, shopping malls involve efforts to create such an experience, and the notion of total immersion in zoo design is to permit the visitors to experience the animals' natural habitat. One can also be immersed in debugging a computer program or repairing a car or producing a showpiece cake.

The sense of being in a whole other world implies extent - either physically or perceptually. To achieve the feeling of extent it is necessary to have interrelatedness of the immediately perceived elements, so that they constitute a portion of some larger whole. Thus there must be sufficient connectedness to make it possible to build a mental map and sufficient scope to make building the map worthwhile (S. Kaplan, 1973, 1978).

The interrelatedness can also be at a more conceptual level, encompassing the imagined as well as the seen. This is often true in theater, but it was equally evident in the case of the wilderness. Thus there is a promise of continuation of the world beyond what is immediately perceived. One can experience extent at an even more abstract level as well. There can be a sense of connectedness between what one is experiencing and what one knows about the world as a whole.

Fascination

In addition to the need for extent, another element missing in many discussions of escape is some source of interest or fascination. A fascinating stimulus is one that calls forth involuntary attention. Thus fascination is important to the restorative experience not only because it attracts people and keeps them from getting bored but also because it allows them to function without having to use directed attention. Despite the apparent significance of fascinating stimuli, there has been little study of what contents have this property (Hidi & Bard, 1986). Some promising candidates would include sex and violence, competition and cooperation. It would also seem appropriate to include many of the objects found in nature. To James's "wild animals" one might be tempted to add sunsets and waterfalls, caves and fires.

Central as such fascinating objects may be to recovering one's capacity for directed attention, they can only be part of a larger picture. Much of human fascination revolves around issues of process as well as content. Humans are fascinated by carrying out various informational activities under circumstances of some uncertainty. They are fascinated by attempting to recognize in instances where recognition is difficult but not impossible. They are also attracted to predictions of uncertain events; gambling provides a classic example. And they are fascinated by learning, by following the thread of something of interest in order gradually to acquire a bigger picture (Mueller, Kennedy, & Tanimoto, 1972), rather than by simply being taught new things. The strong preference for scenes high in mystery (chapter 2) presumably expresses a related concern.

These process fascinations are not, however, engaged merely by random sequences of interesting objects. An occasional fascinating element may challenge one's capacity for recognition, but if unconnected to a larger framework it will be only a momentary diversion or distraction. Even an extended sequence of fascinating elements, if unrelated to each other, will not engage our process fascinations. Thus fascination and extent are mutually supportive. Connectedness, or relatedness, or the existence of some larger pattern is required in order to engage this high-level human motivation.

Action and Compatibility

An environment may offer fascination and extent but still fall short as a setting for restorative experiences. An additional ingredient is the degree of *compatibility* among environmental patterns, the individual's inclinations, and the actions required by the environment (S. Kaplan, 1983).

An individual's decision and actions are determined jointly by the individual's purposes or inclinations and by environmental limitations or demands. Comparably, the cognitive activity that guides action is stimulated by patterns in the environment as well as from within the person. If these functional domains are mutually supportive – if one's purposes fit the demands imposed by the environment, and the environmental patterns that fascinate also provide the information needed for action - compatibility is fostered.

The importance of compatibility in human functioning is perhaps easier to see in its absence, that is, in terms of the costs of incompatibility. To be effective in an environment that undermines compatibility requires considerable mental effort. For example, one frequently encounters situations in which the most striking perceptual information is not the information needed for action. One might be looking for a crucial turn along a strip development dominated by advertising that is large, diverse, and colorful. Or one might be trying to read a difficult text in a library reading room filled with individuals who are socializing. In such cases, the inclination to pay attention to the striking information must be suppressed, and the needed information must be sought. This struggle to remain effective requires considerable directed attention.

The concept of compatibility first arose in the context of the Outdoor Challenge research. Participants reported a "sense of oneness" with the environment (Talbot & S. Kaplan, 1986), thus describing one extreme along the compatibility continuum. Not surprisingly, they found this intense feeling of compatibility to be a special and uniquely valued experience.[6]

An Irish saying perhaps best captures the essence of this concept. "May the wind be always at your back" does not express the wish that one should get all one wants; it means that one pursues one's purposes in an environment that supports one's efforts.

THE NATURE CONTEXT

One can imagine these four properties - being away, extent, fascination, and compatibility - occurring in a wide variety of contexts. A home workshop, a playing field, a performance, even a Main Street might well embody these properties. But does the context make no difference at all? Do these properties incorporate everything that is important about a restorative environment? Intuitively, one would suspect that the context itself matters. People travel great distances to carry out activities that could be done more conveniently if the context were not important. Let us examine what role context might involve and, in particular, why the natural environment might play a special role in the restorative experience.

It would seem that one reasonable source of information on the role of natural environments would be the canoeists, fishermen, and others who have gone to some effort to carry on their pursuits in relatively pristine nature. Many of the studies of outdoor recreation in wilderness and near-wilderness environments, however, have reported rather puzzling results. They tend to emphasize the motivation to escape and even the seeking of power and affiliation as primary concerns.

Fortunately, Fly (1986) has provided a solution to this puzzle. He carefully reviewed a series of seven studies concerning the motivation for participating in outdoor recreation. All of these studies concerned relatively remote, wild settings. In a thoughtful reanalysis of this work, Fly found that, contrary to the conclusion drawn by the authors of these various studies, "escape" was not the most highly rated concern. Rather, and without exception, "experiencing nature" or "enjoying the natural surroundings" received the strongest endorsement. Though one might wonder why the various authors of these works failed to notice the implications of their own results, it is important to remember that at the time this work was carried out there was no conceptual basis for understanding these findings. As Hebb (1951) has pointed out, scientists are at least as concerned with what makes sense as they are with the data per se.

Studies by Shafer and Mietz (1969) and Rossman and Ulehla (1977) also point to the centrality of the concern for experiencing nature in the context of wilderness recreation. In addition, they found that the appreciation of this remote nature has a strong aesthetic component.

A Theoretical Digression

Despite the apparent discrepancy in the wilderness research literature, there is on the whole ample anecdotal and empirical support for the importance of the surrounding environment in contributing to the restorative process. The conceptual basis for understanding this relationship has two complementary aspects. There is both a functional perspective that looks at the *why* and a molecular aspect that focuses on the *how*.

The functional issue centers around a theme that we considered earlier, namely, that it is adaptive for confusion to be painful because tolerating such a state would be highly dangerous. But avoiding confusion sometimes gets in the way of dealing with it. It would be helpful to be able to distinguish environments where tolerating confusion would not be dangerous. In such an environment one could take one's time to deal with confusion without incurring a disadvantage.

Restated in functional terms, the issue is how one might identify environments that are safe and reassuring. Fortunately, this potentially challenging theoretical bridge has already been made. Recall the strong link (discussed in chapter 2) between settings that are highly preferred and the opportunity to function effectively. Substantial empirical work showed that environments that foster a sense of safety and competence, where a quick assessment leads to the judgment that one could readily make one's way and could explore without great risk, were the more preferred. Such settings should make it possible to tolerate, and hence to deal with, a certain amount of internal confusion.

The molecular aspect of all this (which might profitably be skipped by those who do not care for this sort of analysis) is based on an aspect of directed attention that we have already hinted at, namely, that it has an important role in managing the experience of pain.

Directed attention can be thought of as one means of achieving focus in a confusing environment. The achievement of clarity and focus, in turn, can result in a reduction or elimination of the pain generated by uncertainty and confusion. Thus, although directed attention can deal with such pain only indirectly, it can often be quite effective. There is a major difficulty with this solution, however. Because directed attention is a limited resource, depending on it to ward off pain comes at a considerable cost: Clarity achieved through directed attention is likely to be paid for in terms of increased mental fatigue.

What is called for, then, is an alternative approach to dealing with pain and confusion that would not depend on directed attention - and could even allow directed attention to rest. It turns out that an environment that is experienced as pleasurable can provide such an alternative.

The nervous system seems to be structured in such a way that pleasure and pain tend to inhibit each other; thus the experience of pleasure tends to reduce or eliminate pain. The implications of this property of the nervous system are profound. Because contact with pleasurable stimuli can control pain, it should be possible to confront uncertainty and confusion in environments that are experienced as pleasurable. Further, the experience of pleasure should reduce the need for directed attention. Thus this molecular explanation achieves what is desired at the functional level, namely, that environments that are preferred (and hence experienced as safe) permit resting one's directed attention. A preferred environment is thus more likely to be a restorative environment. And since nature plays such a powerful role in what is preferred, in general terms, there is a theoretical basis for expecting natural environments to be restorative.

Consistent with this expectation is the fact that the nature theme arises again and again in the context of restorative experiences. But what of the four basic properties of a restorative experience that we identified earlier? Do these have any special association to natural environments? And is there any basis for identifying nearby as well as remote nature as likely to have restorative properties? Let us consider various types of encounters with nature with these issues in mind.

Being Away

Psychologically, being away implies involving oneself in cognitive content different from the usual. For large numbers of individuals in the developed countries, nature is no longer the usual everyday content. As such, nature meets this criterion with little difficulty. This is not to say that, for a given individual, there are not many other forms of being away. But nature is unusual for filling this criterion on so widespread a basis. Consider how characteristic it is for a trip to a distant place to have as its focus the natural environment.

In the context of the nearby environment nature may seem to be at a substantial disadvantage, since in a literal sense it is not away at all. But the human is a conceptual animal. The experience of being away involves what is going on in the head as well as what is going on in the environment. The distinctiveness and separateness of the experience from the workaday experience may be as important as the literal distance. In other words, even if one goes no farther than one's backyard, making the rounds to find new buds and to be sure that all is well can feel to the gardener like being quite distant from the world of pressures and obligations.

Extent

In the trackless wilderness, extent comes easily. As we saw in the context of the Outdoor Challenge research (chapter 4), in the wilderness extent is achieved at several levels, both by virtue of the scope of the setting and through a sense of relatedness. The most basic requirement for a feeling of extent is an interrelatedness of the immediately perceived elements of the situation so that they constitute a portion of some larger whole. This sense of connectedness is a matter of urgency to the novice and a source of satisfaction to the experienced naturalist. For the novice, it is important to know that the initial fragments of the mental map one is building are reasonably representative of the larger terrain. For the experienced individual, the fitting of new patterns into old knowledge serves both as an affirmation of previous knowledge and as a fresh source of fascination (S. Kaplan & Kaplan, 1982, chapter 4).

Another factor that can contribute to perceived extent is at a more conceptual level, encompassing the imagined as well as the seen. It promises a continuation of the world beyond what is immediately perceived. In the case of wilderness, variety and sheer physical scale contribute to this sense of extent. But even a relatively small natural environment can contain certain physical features that help make it vast conceptually – such as being big enough and complex enough to get lost in and offering numerous possibilities of what one might encounter along the way.

One can experience extent at an even more abstract level. There can be a sense of connectedness between what one is experiencing and what one knows about the world as a whole. This higher-level sense of connectedness is what gives the "other world" a sense of reality. The wilderness experience is real in some other concrete way, as well as in a somewhat more abstract sense. It is real not because it matches one's maps of the everyday world (which of course it does not do) but because it feels real – because it matches some sort of intuition of the way things ought to be, of the way things really are beneath the surface layers of culture and civilization.[7] The woods in which Thoreau built his cabin were not in fact vast; civilization was not that far away. But for him Walden represented a world unto itself, a place in which one might be self-sufficient and encounter a wide variety of meaningful experiences.

In the nearby setting extent can also find various expressions. Scope can be suggested by the park layout or landscape design. More important than size is the sense that there might be more to explore than is immediately evident. Japanese gardens are particularly effective in such suggestion. The sense of miniature, characteristic of many gardens and some back yards, also achieves extent, although in a somewhat different fashion. The miniature creates extent through intensity rather than through the suggestion of distance. In such cases, a "whole little world" may be captured in a small space.

Gardening provides various means of connectedness, thus enhancing the sense of extent. Some may experience in gardening a historical connection, a tie to former times and generations past. Certainly, many gardeners feel a relationship to a force or system that is larger than they are and that is not under human control. Seymour (1979) makes the point quite explicitly in *The self-sufficient gardener:*

> There is no bad season: every season presents the gardener with a challenge and an interest of its own. All weather is good for somebody, or some plant, somewhere. The gardener cannot change these things. He must accept the challenge of learning to understand the seasons and of adapting himself to work within their never-ending cycle.

In the context of the nearby environment, the sense of being away and a conceptual vastness of the experience are often difficult to separate. An ingenious (as well as humorous) understanding of these issues is a key theme in Frisbie's *It's a wise woodsman who knows what's biting him* (1969). He argues, for example, that taking a walking stick and a backpack along on a walk vastly increases its beneficial effect (which he refers to as its effect on "red blood density"). His sense of the possibility of a conceptual transformation of the nearby-nature experience is well expressed in the paragraph that ends his introductory chapter:

> In the following pages I intend to set forth some of the possibilities of the miniaturized adventure. While it is true that I have never discovered a Pole, sailed singlehanded across the Pacific, or lived off the land while dodging head-hunters, I do have some outdoor expertise to share. I know how to become more excited over seeing a thrush in a bush than Stanley was when he found Livingstone. And I can become at least twice as tired from a few hours of hiking or canoeing as Daniel Boone ever got in his whole life.

Fascination

Though nature is hardly unique with respect to fascination, it is certainly well endowed. There is a large supply of fascinating objects, namely, the fellow creatures that share the planet with us. There are also processes that people tend to find fascinating, such as growth, succession, predation, and even survival itself.

Many of the fascinations afforded by the natural setting might be called *soft fascination*. Clouds, sunsets, scenery, the motion of the leaves in a breeze – such patterns readily hold the attention but often in an undramatic fashion. Some fascination is so powerful that one cannot at the same time think of anything else. Soft fascination, by contrast, permits a more reflective mode.

These examples of settings that encourage soft fascination have two aspects in common. First, the involuntary aspect is of only modest strength. Second, there seems to be an important aesthetic component involved. These two themes fit well with the role of aesthetics, or preference, in the control of pain, discussed earlier. In other words, soft fascination may be a mixture of fascination and pleasure, such that any lack of clarity an individual may be experiencing is not necessarily blotted out by distraction but rendered substantially less painful. This would make possible the simultaneous exploration of other thoughts, including confusing material that has been made tolerable by the presence of pleasurable stimuli. The reflection engaged in by the Outdoor Challenge participants is perhaps more understandable given this interpretation of soft fascination.

The most direct evidence for the availability of fascination in the nearby-natural environment comes from the gardening research. The attention-holding power of the garden was one of the most highly rated benefits in both garden studies. The nearby environment is probably particularly strong in soft fascination. The play of light on foliage, the patterns created by long shadows, the different moods of a nature oasis with changes in weather and season, all combine aesthetics and interest in a way that leaves room in the mind for other thoughts as well. Perhaps this very invitation for the mind to wander enhances the being-away aspect, even as one stares out the window into a presumably rather familiar and even ordinary portion of one's environment.

Compatibility

For some reason many people seem to experience nature as particularly high in compatibility. It is as if there were a special resonance between the natural environment and human inclinations. Functioning in a natural setting seems for many people to be less effortful than functioning in more "civilized" settings, even though their familiarity with the latter is far greater. (Cawte, 1967, described individuals with sufficiently severe mental illness that they cannot function in everyday society surviving successfully in the Australian Outback.) It is hard to avoid the conjecture that the fact that humans evolved in environments far more natural than those in which we live now has something to do with this special resonance (Thomas, 1977).

There is no doubt that there is much fear of the natural environment among those unfamiliar with it. But the amount of experience necessary in order to become comfortable seems in many cases far less than would be the case for other settings. Certainly, many of the Outdoor Challenge participants had little or no prior experience with wilderness; nonetheless, the speed of their adaptation to that environment was nothing less than remarkable.

There is, of course, not a single "resonance" involved here but many. There are many patterns of relating to the natural environment that people seem to fall into rather readily. There are the predator role (such as hunting and fishing), the locomotion role (hiking, boating), the domestication-of-the-wild role (gardening, care of pets), the observation of other animals (bird watching, visiting zoos), survival skills (fire building, constructing shelter), and so on. The result of all this is that people often approach natural areas with the purposes that these areas readily fulfill already in mind, thus increasing compatibility. Even individuals not intending such activities often find themselves aligning their purposes in that direction in the presence of nature.

Some people find it hard to get deeply involved in some activity that is "just a game," that has no larger significance or importance. It is as if such people have a higher-level purpose that they implicitly carry with them, which is that they should only do what "matters," what is "worthwhile." As we have seen, the natural environment is particularly interesting in that respect, in that it communicates a sense of reality. When one gathers wood for a fire or hurries to prepare one's shelter before it rains, there is little difficulty in considering these activities as being worthwhile. Likewise, there is a feeling of importance in watering and mulching one's garden to protect the plants from drying out. At the same time the fact that many people could purchase what they grow in their garden means that the survival issue in a literal sense cannot explain this feeling.

There is another basis for suspecting that a special resonance can exist between people and natural environments, leading to an exceptionally high level of compatibility. A dominant theme in modern psychology is an emphasis on the importance of control to mental health and psychological well-being. This theme has even been suggested as an explanation of the impact of wilderness (Newman, 1980). The experience of the Outdoor Challenge program, however, points in just the opposite direction. S. Kaplan and Talbot (1983) suggest that, rather than leading to control, the wilderness experience leads to a sense of awe and wonder and, at the same time, relatedness.

> Not only are such feelings not conducive to a sense of control, they put the whole issue in a new perspective. It is no longer so important to remain in control at all times; in fact, some of the Outdoor Challenge participants come to recognize their concern about control as a costly and disturbing preoccupation. Thus individuals who had spent many of their waking hours struggling to gain, or to maintain, control, now felt that they could relax and pay attention to something other than their immediate circumstances. They discovered unanticipated possibilities within themselves, and found that they could function quite comfor-

tably in a more unassuming fashion as an integral part of a larger whole. . . . They feel a sense of union with something that is lasting, that is of enormous importance, and that they perceive as larger than they are.

These sentiments would not be likely to have surprised Thoreau, although they are not the stuff of which most psychological theories are constructed. On the other hand, their fit with the views of one of the founding fathers of American psychology is quite remarkable. In his *Varieties of religious experience,* William James (1902) defined the life of religion as consisting of "the belief that there is an unseen order, and that our supreme good lies in harmoniously adjusting ourselves thereto."

This sort of profound sense of relatedness is not restricted to the wilder forms of nature. The garden study finding that those who did not rely strictly on chemicals found greater satisfaction in all phases of gardening suggests a parallel feeling of partnership with the larger forces of nature.

Concluding Comments

The natural environment seems to have some special relationship to each of the four factors that are important to a restorative experience. In addition, as we have seen, the natural environment has an aesthetic advantage since such settings are uniformly preferred over many other environments.

An extended encounter with the natural setting may more readily provide restorative benefits, but the nearby-natural setting provides many of the opportunities, though less intensely. A sense of being away, of being related to some larger context, and soft fascination are certainly achievable in the nearby setting. How well compatibility can be incorporated is not yet clear; this is one of the many topics that cry out to be explored. The study of what makes nearby-natural environments more or less restorative has hardly begun. At the same time, enough is known both to indicate the value of creating and enhancing such settings and to provide some focus for the intuitions that many people have about the cultivation and management of the nature nearby.

Thus the many pieces of the puzzle begin to fall into place. The natural environment is often experienced as a preferred or aesthetic environment. This is true both because the content of nature contributes to an aesthetic experience and because the patterns and rhythms of nature contribute to the process aspect of preference as well. Natural settings often contain mystery and often offer coherence; these are the two most powerful of the process predictors.

The benefits that people experience in nature are closely related to these aesthetic factors. First, aesthetic natural environments give pleasure; they are satisfying to experience. Second, such settings support human functioning. They provide a context in which people can manage information effectively; they permit people to move about and to explore with comfort and confidence. And, finally, such environments foster the recovery from mental fatigue. They permit tired individuals to regain effective functioning.

This third benefit of aesthetic natural environments, the restorative function, is itself rich and complex. One can usefully distinguish four different contributions that a restorative experience might make to the recovery of mental effectiveness. These four aspects are interrelated and appear to form a sequence of deepening levels of restorativeness. Each level calls for both increasing amounts of time and increasingly high-quality restorative settings in order to be achieved.

At the first level is the "clearing the head" function. After completing a task there are a variety of cognitive leftovers, miscellaneous bits and pieces still running around in one's head. Thus one may start the next task with something of a deficit, since the residual clutter is likely to be in the way of understanding the new task's requirements and constraints. Sometimes after intense cognitive activity one can close one's eyes and feel the rapid eye movements related to the many pieces of leftover information. The least demanding role of the restorative experience is probably that it allows these distracting fragments to run their course.

A second function of a restorative experience is, not surprisingly, to permit the recovery of directed attention. As we have seen, this is a vital function because so many important cognitive functions require at least some degree of directed attention.

A third function depends upon the cognitive quiet that is fostered by soft fascination. Most of us carry around a cognitive residue of the preceding days, months, and even years. There are, in other words, matters on one's mind that often go unheard. Facing such matters is important not only because they may have functional importance but also because they too can create clutter, an internal noise that will either muddle thoughts about other issues or require considerable directed attention in maintaining focus despite this potential distraction.

The final level of restorativeness is the most demanding of all in terms of both the quality of the environment and the duration required. It is an aspect of the restorative experience we would never have suspected had it not emerged so clearly in our data. And, like so many other surprises we have happened upon in our research, it makes perfectly good sense in retrospect. A deeply restorative experience is likely to include reflections on one's life, on one's priorities and possibilities, on one's actions and one's goals. Here too the functional benefits can be great. Certainly, making a major effort on behalf of a goal one actually does not care about could be a costly error. Yet, if one never checks on what one is doing, such priority distortions could all too easily occur. Perhaps the hazards of "the unexamined life" are functional as well as moral. And perhaps the "sacred grove" mentioned by the ancient philosophers is indeed the ideal setting for carrying out this important reflective activity.

The relationship of people and the natural environment spans a wide range of concerns, from the pragmatic to the spiritual. The restorativeness concept well illustrates this range. On the pragmatic side is the array of health benefits, both mental and physical, that result from restorative experiences. There may also be a substantial impact on what are called human errors, which result when mental fatigue leads to reduced attentiveness and to clouded judgment.

On the spiritual side is the remarkable sense of feeling "at one," a feeling that often – but not exclusively – occurs in natural environments. Although the spiritual does not hold a prominent place in the writings of most psychologists, the concern for meaning, for tranquility, and for relatedness has not gone unnoticed. A striking aspect of this analysis is that the lofty (but frequently neglected) spiritual domain and the mundane (and often also neglected) practical aspects may have much in common. Certainly, in the context of the natural environment these themes converge to a remarkable degree.

As psychologists we have heard but little about gardens, about foliage, about forests and farmland. But our research in this area has brought us in touch with a broad range of individuals for whom these are salient and even, in their own terms, life-saving concerns. Perhaps it is time to recognize this resource officially for what it is, time for governments and mental health professionals and economists to acknowledge what many others have already figured out. It is rare to find an opportunity for such diverse and substantial benefits available at so modest a cost. Perhaps this resource for enhancing health, happiness, and wholeness has been neglected long enough.

NOTES

1. The use here of a proposal for a mechanism that is nearly 100 years old provides some indication of just how far William James was ahead of his field. Two sorts of evidence indicate that this choice is not a capricious one, even if it has not yet gained wide acceptance. First, it is generally acknowledged by researchers studying brain damage that there exists a global inhibitory control mechanism in the prefrontal cortex. Further, damage to this portion of the brain leads to difficulties in "formulating goals, planning how to achieve them, and carrying out the plans effectively. . . . Impairment or loss of these functions compromises a person's capacity to maintain an independent, constructively self-serving, and socially productive life" (Lezak, 1982). These are precisely the functions one would expect of the sort of inhibitory mechanism that James describes. Adopting this proposal also makes possible the explanation of a range of otherwise hard-to-interpret phenomena (cf. n. 3).
2. Being away, however, plays an important role in the recovery from mental fatigue. A persuasive argument has been made for the existence of inhibitory control in the brain that, unlike directed attention, is relatively content-specific (Milner, 1957). Presumably, when one is so tired of some content that one "never wants to see it again," one has fatigued this level of inhibitory control. The critical test comes when one shifts to a new content domain; if the fatigue and inertia disappear in the new context, the deficit must have been at the level of specific content. Not only does fatigue of this content-based

inhibition reduce one's effectiveness in a particular content domain; in time it can indirectly impact directed attention as well. Consider a situation in which one has worked in a given content area until the content-level inhibition is fatigued. One then has two choices. One can either persist or change to another content. Given widespread pressures favoring specialization, diligence, and "time on the job," there is often a high premium on persisting. Doing so can result in a marked decline in focus, since the content-based inhibition is no longer effective. To offset this deficit, considerable support from directed attention is likely to be necessary. Thus a lack of a change of content can inflict a substantial cost. Conversely, "getting away" to a new content brings in fresh content-based inhibition and hence makes less demand on directed attention. Thus events that permit the rest of directed attention are characteristically "away."

3. Most of the studies cited here have been interpreted by their authors in terms of stress. Two aspects of the stress concept, however, are limiting with respect to many of these findings. First, the onset of stress requires an assessment of potential threat or harm. Second, stress expresses itself in terms of increased autonomic nervous system activity. It is not clear how such a state would influence higher cognitive functioning. There is then the further challenge of relating these cognitive deficits to the socially negative inclinations found in such studies. Despite their apparent diversity, these phenomena are readily accounted for in terms of the broad inhibitory deficit that we have called mental fatigue.
4. Portions of this section originally appeared in somewhat different form in S. Kaplan and J. F. Talbot, "Psychological benefits of a wilderness experience," in I. Altman and J. F. Wohlwill (Eds.), *Behavior and the natural environment.* Copyright 1983 by Plenum Press. Adapted by permission.
5. Whereas the Outdoor Challenge research relied heavily on self-report data, subsequent research to test the theory of restorativeness has been more experimental. Mang (1984) demonstrated that a wilderness experience does, in fact, lead to a more rested directed attention capability, as measured by a proofreading task. Hartig, Mang, and Evans's (1987) study provides further support that a nature walk is more restorative than either an urban walk or a rest period (with music and magazines) in a laboratory setting.
6. At first glance the compatibility concept might appear to be merely another name for what Csikszentmihalyi (1975, 1978) has called "flow." Upon closer inspection, however, it is clear that the concepts are quite different. Flow is a state of total involvement. Often the outcome is highly charged. Such examples as surgery and mountain climbing deal with life and death; although the games used as examples (e.g., chess, basketball) do not deal with such dramatic alternatives, winning and losing are taken quite seriously in Western culture.

Consistent with the gravity of the potential outcomes, high feedback is a hallmark of the flow experience. A closely related issue is the presence of a well-defined structure of correct procedures. In games this is quite explicit, but even in surgery and mountain climbing there is a strict body of expectations of what is correct behavior. Most of the examples of flow involve highly skilled individulas who have thoroughly incorporated the set of expectations appropriate to their activity.

Compatibility, by contrast, need not involve drastic outcomes or rules or high feedback. Nor does it require that the individual be highly skilled and highly trained. It is a more democratic state of mind, more readily available to everyone. It is closer to the idea of a good fit than to total involvement. It is hard to imagine an individual in the midst of a flow experience being in a reflective frame of mind. Yet we have found compatibility to be an important ingredient of the reflective mode. Perhaps the brief anecdote related by Watzlawick, Weakland, and Fisch (1974, p. 58) makes the contrast most effectively: "When an eager pupil, in his frantic quest for *satori,* asked the Zen master what enlightenment was like, he answered: 'Coming home and resting comfortably.'" This description hardly applies to flow; it fits compatibility perfectly.

7. Our analysis of this issue parallels that of Brickman (1978), for whom feeling real and having real consequences are important criteria.

CHAPTER 7

THE MONSTER AT THE END OF THE BOOK

IN a project to develop a master plan for a major entryway into a city, a key player has been an influential member of the Chamber of Commerce who feels that the only reasonable policy is to develop all open land that might lead to an increase in commercial activity. He personally, however, prefers to live in the country and looks forward to going home to his boat on the lake by his house. This schizoid perspective well expresses the current perspective on nature in our culture. It is personally valued but carries little clout in the policy arena.

Given the many benefits conferred by contact with natural environments this policy myopia is remarkable indeed. And when one considers the possibility that some of the stubborn social ills that plague our society might be ameliorated by wider access to nature, this bias seems stranger still.

Part of the explanation for this peculiar and maladaptive stance may lie in the traditional hesitancy to exert control over private property for the public good. We have come to see, however, that the economists' faith in the "invisible hand" has failed to protect such common goods as clean air and clean water. Perhaps we will someday have the courage to add natural environments to this list.

A second stumbling block to making protection of natural environments a policy imperative is the scarcity of evidence or documentation of the importance of such settings. Perhaps too there is a lack of adequate conceptualization of how natural environments could possibly have more than a decorative role. For too long nature has been the province of the poet, the artist, and the designer. Poems, paintings, and designs are undoubtedly valuable cultural achievements. Their influence on policy, however, has traditionally been limited. It may be time for science to share some of the responsibility for bringing these powerful potentials to the attention of the "practical" people who make the "hard" decisions.

But, despite the efforts of the kind described in this volume, the decision makers may feel that there are not yet sufficient grounds for action. Though this may be because they are more comfortable with economic projections than with nonnumeric intangibles, perhaps they would find an analogy helpful.

Consider a small industrialized country somewhere faraway. One of the major industries of this country has for years dumped its waste product on the ground. There is now evidence that this material is showing up in groundwater. Further, it is suspected of leading people who drink it to be more irritable (and quicker to violence), less effective (and more likely to attempt to escape problems through drugs rather than to try to deal with them), less able to exercise self-control, plan for the future, and make thoughtful decisions. Let us say further that these people were experiencing an increase in various health symptoms and an overall decline in their health. Should the country move quickly to deal with the situation? Would not prudence dictate that action be taken even if definitive evidence were not yet available?

Our problem is, of course, different in a number of respects. Most importantly, the damage arises from the absence of something (i.e., the natural environment) rather than its presence. This may make the problem harder to grasp, but the issues of urgency and prudence are no less compelling. The decline in access to nature continues unabated in many ways and in many places. At the same time the social ills of person abuse, of drug abuse, and of general malaise resist solution. Even if the natural environment could account for only 10% of these problems, it would seem only prudent to try to claim that 10%.

The "Sesame Street" book, *The monster at the end of this book* (Stone, 1971), presents one of the engaging monsters of "Sesame Street" trying to convince the reader not to turn the page because "there is a monster at the end of this book." He erects a sturdy brick wall to aid in this goal, and when one does turn the page there is the monster sitting in the rubble that was once his brick wall, saying, "I thought I told you not to turn the page." As the book goes on, the monster becomes increasingly frantic in his efforts. Finally one does come to the end and the engaging, helpful monster discovers that *he* is the monster at the end of the book. He is greatly embarrassed.

When we come to the point of asking who is to blame for not preserving the sort of environment in which our species thrives, who has permitted the endless encroachment on the health-restoring presence of the natural world, the answer is very clear. It is we who have allowed this to occur. We are the monster at the end of the book.

Yet the situation is by no means beyond correction. Just as there are many levels at which the natural environment provides a healthful and beneficial environment for people, there are many levels where corrective and protective action can take place. Unlike the solutions for the multitude of the world's current environmental and social ills, many of the needed actions with respect to the natural environment can be achieved affordably. Furthermore, many of them can be accomplished at a local level, with direct benefit to those participating in the solution.

Local actions and solutions that are participatory can, as a matter of fact, be powerful forces in protecting and restoring environments. What appears as vacant land on the land-use map in the planning office is often regarded as a treasured patch of natural environment at the local level. The value the public at large tends to place on nature, however, extends well beyond the local. Benefits are also derived from natural areas that can be visited or seen in passing. There can even be substantial satisfaction from "knowing that it is there." Perhaps for these reasons, when public participation is considered an important part of the decision-making process, the natural environment is likely to benefit. And participation in decision making, though of great value when properly conducted, does not exhaust the participatory possibilities. Although often resisted by administrators, participation in construction and maintenance is often beneficial in terms of cost saving, involvement, and satisfaction.

We have seen the importance of scenery as a natural and psychological resource. We have seen the potential power of wilderness, both in the literal sense and in the broader meaning of opportunities for hiking and camping. And, finally, nearby nature and gardens deserve far more standing than they usually are accorded. Viewed as an amenity, nature may be readily replaced by some greater technological achievement. Viewed as an essential bond between humans and other living things, the natural environment has no substitutes.

APPENDIXES: SUMMARIES OF STUDIES AND PROCEDURES

OVERVIEW OF PREFERENCE RESEARCH METHODOLOGY

There are numerous approaches both to the study of environmental preference and to the problem of data reduction. The purpose of this appendix is not to review either of these literatures. (For an excellent review of methodological issues, see Vining & Stevens, 1986.) Rather, the intention is to describe the procedures that are common to the many studies that form the basis for the discussion in part I and that are summarized in Appendix B. (The same approach to data reduction was also used in many of the studies discussed in part II.)

THE VISUAL ENVIRONMENT

Much of the way humans experience the environment is visual. An understanding of the experience of the environment thus requires visual material. (The issue here is not to exclude verbal expression, but not to rely on it without corroboration from reactions to visual input.)

In the context of research, either the environment can be studied in situ (where it is), or it needs to be represented. To take study participants to the environment has many limitations and precludes study of many settings that are not accessible. One can represent the environment in a variety of ways, and all of us are exposed to these frequently. The approach that is most adaptable to the research context involves the use of pictures – photographs, slides, or drawings.

Fortunately, studies that have investigated the limitations of such surrogates for the actual environment have found that the use of photographs and slides poses no serious problems (Pitt & Zube, 1987). (The reliability of using drawings has received little attention.) All the studies discussed here are based on the use of photographs or slides of existing environments.

ENVIRONMENTAL SAMPLING

Research always involves sampling, but in much social science research sampling issues are limited to the selection of study participants. When studying the environment, however, the question of adequate selection of environments is equally vital. Just as the selection of study participants limits the researcher's ability to generalize the results to other individuals, the generalizability to other settings (or environments) is constrained by the sampling of environments.

The issue of environmental sampling, then, is of major importance to the approach we have taken. If one studies preference for coastal settings, for example, a decision needs to be made about what kinds of coastal settings are to be included. For each identified type, the study must include several instances. If the study were to include two or three examples of coastal scenes, it would be difficult to ascertain whether the preferences were affected by some idiosyncratic aspect of these scenes – for example, the cloud cover or the boats or the water texture. Similarly, if each different type of coastal setting included in the study were represented by a single photograph, one could not determine whether the participants' reactions are to the type or to something particular to the scene.

Determination of the appropriate number of instances and the feasible number of types to include in a study involves a combination of many factors. As is true with many research decisions, there is always an element of arbitrariness. There is also an element of intuition that is the result of experience. Another factor guiding the decision involves concern for what is a manageable task for the participants. A photo-questionnaire sent to people's homes that includes 20 pages of pictures to be rated is unlikely to be returned. Asking participants to sit and look at hundreds of slides also raises problems, even if the psychometric lore dictates that validity is enhanced by having more instances.

We have found four pages of photographs, with eight scenes on each page, to be a comfortable number to include in a mailed photo-questionnaire. Similarly, 60 scenes pose no problems when a group of individuals is asked to view slides and to rate each one. These rough guidelines mean that any single study must confront the problem of how many types of settings can be included and often ends up being of narrower scope than had originally been desired.

In selecting scenes to include, the researcher needs to be articulate about criteria. Photographic imperfections are an obvious reason for excluding scenes. Decisions about detail (close-ups) also need to be considered. Whether or not scenes are to reflect seasonal variation and climatic variation also requires decision. It is not unusual to start with even tenfold the number of scenes that will eventually be included. The more explicit one is in deciding on the types of environments and the selection criteria, the more efficient this process is likely to be.

The richness and usefulness of the preference procedure ultimately depend upon these decisions about the types of environments to be sampled. Content knowledge is helpful here (one's categories are far more likely to be appropriate for familiar environments), as is experience with the procedure and knowledge of the pertinent literature. That is not to say, however, that the choice of types needs to be correct for the outcome to be useful. Selecting a variety of types insures inclusion of a variety of scenes, even if the empirically derived categories are different from the initially designated types.

Though there is no substitute for "knowing the territory," a few guidelines for effective sampling may provide a helpful starting point:

1. Avoid sore thumbs. (If a scene has something visually striking that most other scenes do not have, its other content is likely to be ignored.)
2. Any category sampled should have several scenes (at least three, preferably four or five) representing it.
3. Span the range of preferences available in each category. (Avoid the temptation to use only the more aesthetically pleasing scenes; to do so would be to bias the ratings on the category.)
4. If a scene too important or too interesting to drop could be interpreted in a variety of ways, include several scenes that fit each of the alternative interpretations to permit subsequent hypothesis testing.

Beyond these guidelines it is difficult to have clear rules about environmental sampling. Environments differ, and the constraints of studies lead to many considerations. The purpose of a study necessarily impacts the sampling of scenes. Time and budget limitations also lead to necessary compromises. The most important rule, however, is that attention to environmental sampling is of utmost importance.

THE TASK

All of the studies on which the discussion in part I is based involved asking the participant to rate each scene in terms of preference. Here we are placing no importance on the issue of sampling. In other words, individuals are not asked how much they like the scene, how pretty they think it is, how they would judge its scenic quality, how much they would like to go to such a setting, whether they would like to own the picture, and so on. Sampling the domain of preference is far less important in our approach to this research area than is sampling the environment.

The reason for this is straightforward. Although the responses to these different domains of preference may be somewhat different, the pattern of responses will be very similar. A scene one likes more will be considered more beautiful and higher in scenic quality. Not surprisingly, when studies include several such tasks, the correlations are extremely high (Pitt & Zube, 1979). The emphasis in semantic differential approaches to preference research is similar. The consistent finding that the adjectives related to preference group into an "evaluative" dimension tells us more about how people use language than about how they experience space (R. Kaplan, 1975*).

Whether one asks participants to use a 5-point rating scale or to place the scene in one of five piles or to circle a word or phrase selected to represent each of the points on the scale ("not at all," "somewhat") makes little difference in terms of the results. (The choice of approach can, of course, be important for the context of the study [R. Kaplan, 1984d].) The number of scale positions also is somewhat flexible. Permitting the participant a neutral midpoint (odd number of scale positions) or forcing a direction (even number of scale positions) also has its adherents. We have settled on five points (from "not at all" to "a great deal"), and this has worked well. (Stino, 1983*, found that her Egyptian participants could relate far more easily to a 10-point scale, which was used in that culture for other situations.)

Many other forms of rating scales have also been used in environmental perception and preference research. Schroeder (1984) provides a careful analysis of the differences among such approaches and concludes that simpler methods have distinct advantages and are "appropriate for a wide range of applications."

CATEGORY-IDENTIFYING METHODOLOGY (CIM)

Conscientious environmental sampling leads to an abundance of data. That can be both exciting and overwhelming. One approach to handling preference ratings for many scenes is to compute the mean rating for each scene. Analysis of the most and least preferred scenes can be very useful (as discussed in chapter 2).

Quite aside from an analysis based on the magnitude of preference (how much each scene is liked) are procedures that involve patterns among the ratings. Are there similarities in the ways some scenes are rated, whether or not they are highly favored? There are dozens of data reduction procedures whose objective is such a search for common patterns. They go by many names and have many mathematical variations in terms of assumptions and procedures.

Most of these approaches begin from the notion of correlation – the relationship between ratings of all pairs of scenes included in the study. If a correlation between two scenes is relatively high (close to 1.0), then if one knows how much one of the scenes is liked one can predict fairly accurately the extent of preference of the other scene. If a study includes, let us say, 40 scenes, then there are 760 such correlations to examine. Such a task is far more overwhelming than exciting.

In this day of fast computation the task of calculating so many correlations is trivial, and the subsequent steps in the data reduction process are also accomplished in a matter of seconds. In order to gain a better understanding, however, it is useful to take a closer look at the steps involved in a CIM procedure.

Let us say our study included 40 scenes of different natural settings in a residential context. Each of 200 participants rated each of these scenes using a 5-point scale. We can array this information in a table where each column is a scene and each row is a person. There will be 40 × 200 numbers in the table. Finding patterns in these numbers is well beyond one's patience, and so correlations have been calculated. These are now arrayed in a table that shows the relationship between ratings (across the 200 participants) for each scene with each other. This requires a 40 × 40 matrix, where the diagonal shows the perfect correlation of each item with itself and the triangles formed on either side of the diagonal are identical. This then has 760 different correlations.

One way to tackle such an array is to decide to examine only the "big" correlations – perhaps (arbitrarily) those greater than .50. So one circles colorfully such values and finds that, rather than having 760 entries to deal with, the "big" ones are few and far between. These now become the set of numbers to examine, and this task is more manageable. One can, for example, look for the biggest values in the table and then look to see what other scenes relate highly to each of these. Perhaps some of these also relate to each other, forming a highly interrelated set of scenes. This is what a *category* means – and the informal procedure we have described achieved *category identification*.

A few points are important to make here. First, notice that a great deal of the information that was originally available in the 40 × 200 table has been lost. We no longer know how any particular individual rated any particular scene. We also dropped from the analysis many scenes that did not relate highly to any others. That's what *data reduction* means – one has reduced the amount of data to be considered in favor of making sense of the material.

Second, the criterion for what correlations to circle affects the final outcome. The number of scenes in a category, the number of categories, the interrelatedness among categories are all affected by such criteria. Each statistical procedure uses different algorithms (and hence different criteria), and these, therefore, affect the results.

Third, the statistical procedures identify categories or patterns. In other words, the final tables produced by the computer consist of values that indicate how strongly each item "belongs" in a category. The computer does not know what the items are. It does not know who the participants were. It does not know that the data are about photographs as opposed to verbal items or geographic regions.

Fourth, the researcher must always interpret these results. The task of naming a category - of finding what its common theme might be - is done by a human, not a machine. The preparation of material to make it more interpretable by the human eye and brain is a trend in other sciences as well. In this way the strengths of both computers and humans are most effectively utilized.

SPECIFIC CIM PROCEDURES

Most of the studies discussed in part I relied upon two specific forms of CIM, ICLUST, and SSA, and in most cases both procedures have been used with the same data.

ICLUST (Kulik, Revelle, & Kulik, 1970) is a hierarchical cluster analysis procedure. It transposes the original correlations to proximity coefficients (technically, coefficients of collinearity, which measure the extent to which two variables have the same correlations with all the other variables in the matrix). ICLUST looks for the two most similar items and then considers these a group. Other groups are formed in similar fashion, consisting in each case of either a pair of items, a pair of previously formed groups, or an item and a group. The algorithm continues to form such pairings until each item has been joined in a group, as long as such combinations yield higher alpha coefficients (a measure of internal consistency). The result is a tree structure where each scene is uniquely placed in a group. Among the advantages of this procedure are its efficiency, its spatial display, and that it computes the coefficient of internal consistency.

Mathematically, the Guttman-Lingoes Smallest Space Analysis (SSA-III) is a totally different procedure. It is a nonmetric factor analytic procedure, described by Lingoes (1972) as a "monotone vector analysis." At an intuitive level, SSA-III transforms the correlation matrix to a rank-order matrix, thus "throwing away" the specific metric information of the actual magnitudes of the correlations among the items. The effect of this transformation (in other words, the nonmetric aspect of the analysis), in our experience, is to increase greatly the stability of the solution. When the same items (scenes, verbal items, etc.) are used with different samples of participants, the categories that emerge show marked consistency.

As with any factor analysis, the procedure yields a matrix that indicates the degree to which each item belongs to each category. The technical expression for this notion of fit or belonging is *loading*. This is the correlation of the item with the category. We have adopted a loading value of .40 as the threshold: Items with loadings that are above this value define the composition of a category. There are three other criteria, however, that may lead to the elimination of items despite their having a loading of .40.

1. Eigenvalues must be greater than 1.00. This is a widely adopted convention, referring to the total explanatory power of the category.
2. To be designated a category, there must be at least two items. (In some studies, we have used a criterion of three items.) The rationale here is the same as the environmental sampling issue. If a category consists of too few items, it is difficult to ascertain what it is about.
3. An item that has a high loading (>.40) on more than one category is not included in the definition of any of the categories. If such items were included with each of the categories to which they belong, there would be a necessary relationship between the categories. By their exclusion, the categories can be more clearly independent. In the case of ICLUST such duplication of inclusion cannot occur as each item is uniquely placed in the tree structure. This is both helpful and somewhat misleading. The SSA solution helps one see that such purity of categorization may not exist.

CONCLUDING COMMENTS

The methodology described here is not merely a set of procedures but rather an expression of an attitude and approach to doing science. It is an approach committed to conceptualization and to discovery; it is both systematic and flexible (S. Kaplan & Kaplan, 1982, chap. 9). The more thought one gives to the conceptualization of the visual environment, the more one becomes cognizant of its multidimensionality, its amorphous nature. To measure such complexity adequately is likely to lead to many insights.

The use of multiple analytic procedures as well as the insistence on environmental sampling reflect the realization that one's conclusions are necessarily a result of one's procedures. These procedures invite tension: The different CIM approaches may lead to different categories; the clear and simple notion one held about the environment may have to be replaced with something far more complex. Such tension can be viewed as problematic or as a form of discovery.

Unlike procedures where the outcome is yes or no, significant or not significant, here the investigator is confronted with a display, not a binary result. Thus the process is at least as much one of hypothesis generation as of hypothesis testing. One faces a set of empirically determined visual categories, a circumstance that both stimulates and challenges what Gregory (1966) has referred to as "eye and brain." It fosters a more organized and meaningful view of the world, a new way of seeing. As such it is a step toward expertise. It also leads to a sense of greater understanding. Novice users of the technique, in fact, frequently express astonishment at its insight-fostering power.

APPENDIX B

PREFERENCE STUDIES
ABSTRACTS OF STUDIES FOR PART I

These studies form the basis for the analyses presented in chapters 1–3. They are marked with * when mentioned in the text.

The following studies are included in Appendix B:

Anderson, 1978
Ellsworth, 1982
Gallagher, 1977
Gimblett, Itami, & Fitzgibbon, 1985
Hammitt, 1978
Herbert, 1981
Herzog, 1984
Herzog, 1985
Herzog, 1987
Herzog, 1989
Herzog, Kaplan, & Kaplan, 1976
Herzog, Kaplan, & Kaplan, 1982
Hudspeth, 1982
R. Kaplan, 1973b
R. Kaplan, 1975 (included with S. Kaplan et al., 1972)
R. Kaplan, 1977a (Swift Run Drain Study)
R. Kaplan, 1977a (scenic roadsides)
R. Kaplan, 1984b
R. Kaplan, 1985a
R. Kaplan & Herbert, 1987
R. Kaplan & Herbert, 1988
R. Kaplan, Kaplan, & Brown, in press
R. Kaplan & Talbot, 1988
S. Kaplan, Kaplan, & Wendt, 1972
Levin, 1977
Medina, 1983
Miller, 1984
Stino, 1983
Ulrich, 1974
Weber, 1980
Woodcock, 1982
Yang, 1988

Eddie Anderson (1978)
Doctoral dissertation, University of Michigan
(Rachel Kaplan & William F. Bentley, cochairpersons)

Overview. Anderson examined visual preference for forest harvest and regeneration methods and land management practices. In light of the differences in visual preference between foresters and locals and along racial lines, he makes suggestions for integrating the visual with other resource considerations and discusses the implications for agencies involved in public participation, landscape design, and ways to improve visual resource perception.

Setting/Photographs. The photographs were from forest areas in Lake County, Michigan. Criteria for the 48 pictures included being at eye level; reducing "ephemeral" effects such as clouds, fog, and haze; using no people, water, or close-ups. All were taken at the height of the growing season and are representative of the vegetation, landscapes, and forest management techniques (two photographs included two practices not used in the area).

Ratings. A jury of natural resource professionals judged each photograph for the extent to which it had five preference predictors (Legibility, Coherence, Spaciousness, Complexity, and Mystery). A second jury of foresters rated each according to the forestry practices it showed.

Survey/Sample. In addition to rating each photograph on a 5-point preference scale, participants answered a four-page questionnaire (using 1–5 scales) on their perceptions of concerns, activities, and problems of those living in a rural forested community. The 300 participants included 27 resource professionals, 195 residents, and 78 students from Lake County.

Definition of Predictor Variables

Coherence – the consistency of relationships and features in the landscape as they fit together to form recognizable patterns
Complexity – the frequency, diversity, and arrangement of factors
Legibility – confidence in maintaining orientation; predicted ease of traversing beyond the visible ground plane

Mystery - new information suggested but hidden from view; sense of being drawn or pulled into the scene
Spaciousness - perceived or experienced depth

Results

Categories. Photo clusters using ICLUST and SSA-III produced five categories: Heavily Manipulated Landscapes (recent clearcuts, cutovers), Dense Forest and Modest Harvest Practices (older regenerated stands), Pine Plantations, Planned Spacious Openings and Scenic Roads (roads bounded by pole and sapling oak, picnic and camping grounds), Open, Unused Land (wildlife openings, meadows, old uncultivated fields). Table B.1 shows mean preference ratings for each category as well as jury-based evaluations on the predictor variables.

A panel of six forestry professionals was asked to group the scenes according to harvest/silviculture practices. Of the 16 groupings they generated, 4 corresponded to the empirical findings (plantations, recreational areas and roads, and openings). The other 12 fell into three different groupings - recent-generation cuttings, mature growth/regeneration, and managed/unmanaged oak.

Predictors of Visual Preference. Although Coherence, Mystery, Complexity, and Spaciousness proved the best predictors of preference, there were significant differences among the sample when other variables were examined: familiarity (length of exposure), race (black and white), level of expertise (whether or not one was a natural resource professional). Although Coherence and Complexity were the best predictors of expert preference, Coherence and Mystery best predicted the preferences of students and residents. Spaciousness and Legibility negatively

Table B.1. *Category means and predictor ratings (Anderson, 1978)*

Category	Mean pref.	Comp.	Myst.	Coh.	Legib.	Spac.
Heavily Manip.	2.31	m	l	m	m	h
Open, Unused	3.22	m	l	h	h	h
Dense Forest	3.55	m	m	m	l	l
Pine Plantations	3.66	l	l	h	m	l
Pl. Spacious	3.94	h	m	h	h	h

Note: l = low; m = medium; h = high.

predicted preference, with Spaciousness playing a more significant role than Legibility only for black students and professionals.

Preference. The results showed that professionals (experts) responded differently from both residents and students. They had significantly higher preferences for Heavily Manipulated Landscapes (3.4–2.2/2.3, respectively), Dense Forest (3.9–3.5/3.5) and Open, Unused Land (4.0–3.1/3.3). There were no significant differences between ratings of Red Pine Plantations (3.9/3.2/3.5) or Planned Spacious Openings (3.8/3.9/3.8). Experts also showed less sensitivity to the range of scenes.

With increased familiarity (measured by length of residence and exposure), preference among experts for Dense Forest was higher, and was lower for Heavily Manipulated Landscapes and Open, Unused Land. Student preference for Heavily Manipulated Landscapes also was lower. Resident preference for Open, Unused Land was higher.

For the most part, students and residents showed a high level of consistency, but there were some racial differences. Blacks had a significantly higher preference for Planned Spacious Openings whereas whites had a higher preference for Dense Forest. White students liked Heavily Manipulated Landscapes, Dense Forest, and Open, Unused Land significantly more than did black students.

The questionnaire responses clustered into 10 categories, which were further collapsed into two: Community-Based Preferences and Forest and Out-of-Door Activities. Findings supported the expert/citizen differences suggested above. Professionals were more positive about the people, community, and forests of the area, about their own career fields as their lifetime vocations, and about the satisfaction they draw from solitude and outdoor stress releasers (although these activities were the most highly rated by all participants). Residents and students found their home and community activities affording them more satisfaction than did the professionals.

Noteworthy Points

Experts and their different preferences and perceptions of visual landscapes
Role of familiarity in visual preference
Racial differences

The dissertation also includes an extensive overview and analysis of recognized approaches to the visual preference literature and empirical studies up to 1978.

VISUAL ASSESSMENT OF RIVERS AND MARSHES: AN EXAMINATION OF THE RELATIONSHIP OF VISUAL UNITS, PERCEPTUAL VARIABLES, AND PREFERENCE

John C. Ellsworth (1982)
Master's thesis, Utah State University (Craig Johnson, chairperson)

Overview. Ellsworth compared the effectiveness of two visual methods in predicting preference. One method uses a descriptive inventory of biophysical features and professional aesthetic evaluations; the other, variables reflecting people's perceptions of landscape (Legibility, Coherence, Complexity, and Mystery). The research also provides preference data on rivers and marshes, two settings that have not been looked at in other empirical visual studies.

Site Visual Assessment Methods

Visual Unit Data Collection. Three levels of visual units were identified in a marsh area and river tributaries of the Cutler Reservoir, Utah. The three levels - landscape, setting, and waterscape - were based on evaluations of biophysical similarities and consistencies in water expression (e.g., open, marsh, stream), vegetation type, edge condition, and human use and impact.

Photographs and Perceptual Variables. Four groups of judges (18 landscape architecture students) individually rated 76 color slides of the areas mentioned above in terms of the degree of expression of one of the perceptual variables: Coherence, Legibility, Complexity, and Mystery.

Sample/Survey. Ninety-eight college students divided into two groups, viewed 60 of the above 76 slides (10 representing four river waterscape units, 20 of the marsh setting unit). The order of slides was counterbalanced for the two groups. Respondents rated each scene on a 5-point scale in terms of how much they liked it.

Definition of Predictor Variables

Coherence – extent to which the scene "hangs together" (redundant elements, textures, and structural features allow prediction from one portion of scene to another; organization causes elements to be perceived as groups).
Complexity – number of visual elements in scene; how much is going on.
Legibility – prediction of the opportunity to function; finding one's way in, and finding one's way back; ease of forming a "mental map."
Mystery – going into the scene seems likely to provide more information (it must appear possible to enter scene and go somewhere; promise of further information based on a change of vantage point).

Results

Preference. Differences in preferences for the five visual units were significant only between the river and marsh unit settings (combined river mean pref = 3.06; marsh = 2.47). All river scenes rated higher in preference than the marsh scenes. The most preferred river scenes were biophysically diverse and high in Mystery and visual depth. Respondents preferred pastoral settings with curved stream corridors over those with trash, unclear spatial definition, and obscured views. Marsh images with Mystery and depth were most preferred, especially those with foreground vegetation and clear water reflections. Demographic variables did not significantly affect preference.

Perceptual Variables. Correlations of the perceptual variables suggest that they are independent of each other. Significant correlations between the variables were: Coherence and Complexity, $r = -.51$; Coherence and Mystery, $r = -.41$; and Complexity and Legibility, $r = -.27$. The Coherence/Complexity relationship was influenced most by the marsh setting data ($r = -.75$).

Three of the variables significantly correlated with preference: Coherence, $r = .33$; Complexity, $r = .37$; and Mystery, $r = .41$. The relation of Complexity to preference was especially high in the river scenes, suggesting that visual diversity is an important predictor of preference. The relationships of Mystery and Coherence to preference were strongly influenced by the marsh values ($r = .50$ and $r = -.44$, respectively). A highly coherent marsh scene was often visually simple and monotonous (therefore the negative relationship between Coherence and Complexity and between Coherence and preference).

Visual Units versus Perceptual Variables and Relation to Preference. Results from regression analysis showed that perceptual variables explained more of the variability in mean preference than did visual units ($F = 2.54$, $df = 4, 51$).

Noteworthy Points

Effort to relate and compare two visual assessment approaches (visual units and perceptual variables); suggestion that preference is not as related to visual units as has been assumed by other researchers

Importance of "involvement" components (Complexity and Mystery) in preference

Water is known to have a positive influence on preference, therefore, "lower" preference for marsh is relative within the context of this study and not directly comparable to other studies

The *content* of human influence is an important factor in preference; those scenes with trash and disorderly elements were much less preferred than those depicting a nearer, more orderly relationship

Ed. note. CIM-based analyses of these data were not included in the thesis but are included in the analyses in the text.

VISUAL PREFERENCE FOR ALTERNATIVE NATURAL LANDSCAPES

Thomas J. Gallagher (1977)
Doctoral dissertation, University of Michigan (Rachel Kaplan & S. Ross Tocher, cochairpersons)

Overview. Gallagher examined people's visual preference for more natural, lower-maintenance designs within the urban setting. His goals were to identify strategies designers can use in design or in working with people to gain acceptance of natural landscape proposals, to find the relation between design variables and visual preference, and to discover other predictors of preference.

Setting/Photographs. The photographs were of 32 scenes of the CUNA insurance office property in Madison, Wisconsin, including both a prairie/woodland and traditional ornamental landscaping. The site was chosen because of efforts to incorporate a natural landscape (prairie/woodland) in a traditional manicured lawn setting.

Ratings. A jury selected photographs for their understandability, photographic quality, normal framing and composition, ordinariness, and representativeness in terms of available scenes. A second jury then rated the photographs for seven preference predictors (see Table B.2; The variable Trees was suggested by preliminary results and was juried later).

Survey Sample. The survey involved preference ratings (1–5 scale) of each photograph. Other measures dealing with attitudes and beliefs about the area's appearance, problems associated with natural landscapes, and descriptions of the areas were rated on a 1–6 disagree/agree scale. The sample included 137 employees (both staff and professionals) of the company that owned the grounds and 36 residents living near the property. The majority were white, middle-class. Surveys were distributed to all those homes abutting the property or separated from it by one house, to a random selection of residences more than two houses from the property, and to 13 randomly selected apartment dwellers. Of these, 65% responded to the survey: 95% of those living nearest, 20% of the more distant group, and 65% of the apartment dwellers.

Definition of Predictor Variables

Coherence - how well scene "hangs together"; how easy to predict from one portion of scene to another

Complexity - extent to which scene presents many visual elements or promises more information if only one had more time to look at it from present vantage point

Edge - presence of distinct borders between zones; legibility through order

Mystery - extent to which scene promises more information if viewer could walk deeper into it

Naturalness - opposite of ornamental

Spaciousness - visible availability of options for locomotions; how much room to wander

Texture - how fine-grained ground surface is

Trees - number, size, and dominance in scene

 Results

Categories. The photographs fell into five themes using the ICLUST and SSA-III procedures. Prairie (tall grass, dense shrubs, no trees or buildings), Building (building, prairie, lawn), Complex (abundance of trees), Simple (ornamental landscaping, some trees), and Lawn (lawn and building, no trees). (Table B.2 shows mean ratings for each category, as well as jury evaluations on the predictor variables.)

Predictors of Visual Preference. Mystery, Coherence, and Naturalness were the strongest predictors. Gallagher concludes that the relationship between Naturalness and preference "suggests that there is a great deal more variability in preference for 'natural landscapes' than had been proposed in past studies where 'natural' was contrasted to 'urban'" (p. 42). Gallagher posits a curvilinear relationship between preference and Naturalness (on a continuum from nature to ornamental or manicured–urban). Given the high negative correlation ($r = -.88$) of Naturalness and Coherence, Gallagher suggests it may be the lack of Legibility rather than increased nature content that accounts for this continuum. Mystery was higher than the more frequently cited Complexity as a predictor of preference. Gallagher remarked on the unexpected lack of significance of Texture and Spaciousness as predictors but noted these predictors were highly intercorrelated with Coherence and less so with Edge.

The strongest combination of predictors emerged with the incorporation of a new predictor, the presence or absence of trees. Trees correlated strongly with Mystery and

Table B.2. *Category means and predictor ratings (Gallagher, 1977)*

		Promised Info		Legibility		Content	
Category	Mean pref.	Comp.	Myst.	Coh./Spac./Text	Edge[a]	Nat.	Trees[a]
Prairie	2.52	l	l	vl	l	h	vl
Building	2.87	l	l	l	l	l	vl
Lawn	3.47	l	l	vh	l	vl	vl
Simple	3.69	l	h	h	h	vl	vl
Complex	3.97	h	h	l	l	l	h

Note: Jury evaluation of predictors: very low (vl) to very high (vh).
[a]Trees and Edge also associated with Promised Information.

Complexity and substantially reduced their role as predictors. Trees, Naturalness, Spaciousness, and Texture accounted for .81 (R^2 value) of the preference score variance. Trees was also the most powerful predictor of preference, with increased size and number of trees enhancing preference. Gallagher felt this predictor was especially significant because the photographs spanned the possible range from prairie (no trees) to densely wooded areas. (Based on remaining predictors, R^2 was only .20.)

Other variables examined were familiarity, one's relationship to the landscape (whether working or living near it), and knowledge about the natural landscape project, as well as differences between those preferring Prairie and those preferring Lawn. Staff preferred Lawn more and Prairie and Building less than professionals. Those who knew of the intent of the prairie woodland design effort preferred Prairie and Building more and Lawn less. The informed group related more to the Promised Information categories, the uninformed to Legibility components.

Apartment and home residents showed notable differences in preference. Apartment dwellers had substantially higher preferences for Prairie and lower preferences for Lawn. For this group, Legibility predictors were negatively and Naturalness was positively correlated with preference. Spaciousness and Texture emerged as strong negative predictors.

Noteworthy Points

Content component of predictors (Trees here, but Gallagher also expects water, wildlife, etc.)

Familiarity and involvement with landscape as preference predictors

The attitude portion of the survey emphasized that concerns about the landscape emerge not so much from the design concept or appropriateness of the landscaping as from how it *looks*

MYSTERY IN AN INFORMATION-PROCESSING MODEL OF LANDSCAPE PREFERENCE

Randy H. Gimblett, Robert M. Itami, & J. E. Fitzgibbon (1985)
Landscape Journal, 4(2), 87–95

Overview. Part of an ongoing effort to improve the Brown/Itami model assessing scenic quality, this study focused on the concept of mystery as an informational factor in environmental preference. The study found that people perceive mystery as a distinct landscape attribute and identified five physical landscape factors associated with differences in mystery ratings. These features contribute to mystery to the degree that they promise more information and/or afford a way to explore that promise.

Site Photographs/Sample. On a 5-point scale, 191 black-and-white photographs representative of the diverse land forms and land cover in rural Ontario were rated for mystery. Mystery was defined as "the degree to which you can gain more information by proceeding further into the scene." The participants were 36 landscape architect students who viewed the photographs in three sessions.

Results. A multidimensional scaling procedure (ALSCAL) was used to look at the degree of similarity in the assessment of mystery. The high correlation (.985) and low stress factor (.125) indicate that mystery, as defined in the model, is perceived as a clear, meaningful concept. This procedure also provided an interval scaling of the photographs, so that attributes of scenes rating high in mystery could be analyzed. Assessment of the photographs' composition produced five physical attributes associated with mystery.

Screening - degree to which the larger landscape is visually obscured by vegetation and shading
Distance of View - distance from viewer to nearest forest stand
Spatial Definition - degree to which physical elements enclose the observer
Physical Accessibility - apparent means of moving through or into the landscape
Radiant Forest - where foreground is in shadow and area beyond is brightly lit

Noteworthy Points

Support of validity of mystery in information-processing model
Unlike other studies abstracted here, this one did not entail preference ratings
Its focus is on an analysis of the attributes of mystery in a natural setting

VISUAL AND USER PREFERENCE FOR A BOG ENVIRONMENT

William E. Hammit (1978)
Doctoral dissertation, University of Michigan (Burton Barnes & Rachel Kaplan, cochairpersons)

Overview. Hammit examined the recreational functions of the bog environment. Because it is highly preferred and experienced primarily visually, the bog lent itself well to a study of visual preference and the roles played by familiarity and visitor characteristics. Finding that those areas most preferred are not necessarily those highlighted for their "ecological" interest, Hammit presented the option of using an information-processing approach to site planning and interpretive program design.

Setting/Photographs. The Cranberry Glades Botanical Area, West Virginia, was selected as a prototypical southern bog. Because of the fragile ecological nature of bogs, the 27,000 annual visitors view it from a boardwalk that makes a loop through the area. The 48 black-and-white photographs for the survey were taken by visitors earlier in the season, the U.S. Forest Service, and the researcher. They depicted various habitats and landscapes of the bog, including some scenes that one could not see from the boardwalk, as well as a few scenes from a bog in Michigan. Selection criteria included using scenes most frequently photographed by visitors and those representative of distant, intermediate, and immediate aspects of those scenes. They included no close-ups or people.

Survey/Sample. The pre/post test preference survey used two photo sets, P1 and P2. Sixteen of the 24 photographs were used in both sets. Group 1 took the pretest before walking through the bog; all posttests occurred as participants were leaving the area. Preference was rated on a 1–5 scale.

Group	*n*	Pre	Post
1	160	P1	P2
2	120	-	P1
3	120	-	P2

In the posttest, the 400 interviewees indicated how familiar the photographs seemed (1–3 scale) and provided information on why they had visited, their present and past experience with bogs, and demographic data. The sampling schedule consisted of three periods during July and August 1977. Twenty days were selected during these periods, and 20 interviews were conducted on each of those days.

Results

Bog Preference Categories. Although response to the bog was generally fairly positive with overall mean ratings of 3.90, there were definite preferences for scenes characterized by landmarks, mystery, or an overview. The four categories generated by SSA-III and ICLUST analyses were Boardwalk (mean preference 4.12), Feature, (having identifiable features or landmarks, mean 3.99), Edge (areas where two types of vegetation or zones met, mean 3.77), and Bogmat (mean 3.69). Hammitt noted that in environments as repetitious and nonlegible as a bog, "sense of place" and focus as found in Boardwalk and Feature are especially important.

Familiarity as a Predictor of Preference. Familiarity was examined in terms of (1) participants' rating of scenes that were most and least remembered (on-site memory); (2) the influence of the on-site experience and prior information (the pre/post Group 1); and (3) longer-term familiarity in terms of previous experience with bogs.

On-site Memory. Analysis of the P2 photo set for the most and least remembered scenes showed that people were very cognizant of what they had or had not seen on their walk through the bog. (Agreement between the two P2 groups was rho = .97.) Scenes with aspects of Boardwalk or Features were among the most familiar, whereas those with Bogmat or Edge were the least. These least familiar scenes, however, were also ones people were less likely to have seen, being of the Michigan bog or of places not visible from the boardwalk. Preference and rated familiarity were positively correlated ($r = .53$).

On-site Experience. Comparison of pre- and posttest scores for P1 were used to determine the influence of the on-site experience. The ratings were sigificantly higher when made after the site experience, although the pattern of ranking was much the same for both groups (rho = .86). One interesting result was the markedly higher preference for a scene of an upturned dead tree that received a relatively low rating by the pretest group. The on-site interpretive display explaining the tree's ecological value may have contributed to this striking difference.

Prior Photographic Information. Posttest preferences of those who had taken the pretest (i.e., seen photographs prior to on-site hike) were lower than the posttest-only responses although, again, the rank pattern remained much the same. This "dampening" suggested that the pre/post participants continued to see the photographs as photographs rather than as actual scenes. It is possible that viewing the photographs prior to the boardwalk hike led to unfulfilled expectations.

Previous Visits (Long-term Familiarity). The number of previous visits positively influenced visual preference. The reasons cited by return visitors (nature-related, photography-related) indicated that they perceived different and additional information than the first-time visitor. Preference for all four features increased with the number of visits, with preference for the Boardwalk component related more to social reasons. Higher ratings for the others involved nature and photography-related activities. Both prior information and number of prior visits had an interactive effect on preference. Viewing photographs before the on-site experience dampened preferences of first-time visitors and increased preference of those returning.

Noteworthy Points. Although intuitively appealing to resource managers as a desirable viewing feature, Edge was low in preference. This discrepancy highlights the importance of considering the array of qualities influencing preference; Edge here involved a lack of focus and low coherence.

Examination of different aspects of familiarity
Suggestions for design and educational programs that emphasize information processes
Suggested potential of interpretive materials on preference

Ed. note. Portions of this study have appeared in the following publications:

Hammitt, W. E. (1979). Measuring familiarity for natural environments through visual images. In *Proceedings of Our National Landscape Conference.* USDA Forest Service General Technical Report PSW-35.
Hammitt, W. E. (1981). The familiarity-preference component of on-site recreational experiences. *Leisure Sciences, 4,* 177–193.

VISUAL RESOURCE ANALYSIS: PREDICTION AND PREFERENCE IN OAKLAND COUNTY, MICHIGAN

Eugene J. Herbert (1981)
Master's thesis, University of Michigan (Terry Brown, chairperson)

Overview. Herbert tested a visual resource evaluation model's (Brown/Itami) ability to predict visual preference. The computer model, based on psychological principles and empirically derived preference predictors, used land-use and land-cover characteristics to classify and evaluate scenic quality. The predictions of the model (the "expert" evaluation) reasonably correlated to "nonexpert" preferences. To correct for times when the model did not predict as well, Herbert suggests incorporating the preference predictors Complexity and Coherence into the land-use dimensions.

Site Photographs. The Brown/Itami model was applied to a computerized data base classifying a portion of Oakland County, Michigan, in terms of land-use and land-form characteristics. The dimensions relating to land-use quality were naturalness, land-use compatibility, height contrast; relating to land-form quality were slope/relative relief, contrast, and spatial diversity. A series of matrix commands combined and rated the dimensions for both land form and land use into "equivalence" classes (from low to very high) and plotted them onto the computer map of the study area. This composite map of visual resource quality was transferred onto a topo map with 7,260 cells. Each cell was assigned to an equivalence class; only those classes with 10+ cells were included in the photographic survey.

Fifty-five of the 110 color slides (10 cells of each of the 11 equivalence classes) taken constituted the final photographic sample. Photographed in September before the influence of fall colors, the slides avoided man-made influences (e.g., golf courses, telephone lines), and tried to represent the diversity of scenes within each class. They ranged from rural residential to open fields and dense woods.

Sample/Survey. The 97 respondents were University of Michigan psychology students (45 men, 52 women). To avoid order effects, the slides were presented in two orders. Respondents viewed each slide for 10 sec and rated how much they liked it on a 5-point scale.

Results

Respondent Preference. Preference ratings for the scenes ranged from 1.92 to 4.41. Of the six most preferred scenes, five included water. All were characterized by spaciousness and orderly nature. The sixth depicted a meticulously landscaped residential scene. The six least preferred were predominantly "man-influenced" and looked disorderly and unkempt. Correlations to the model prediction were "reasonable."

Based on SSA-III and ICLUST, four categories were formed, depicting different degrees of nature and human influence.

Manicured Landscapes (3.94, obviously man-modified)
Vegetation (3.58, dense forests to partly open fields)
Pastoral (3.09, open fields and panoramas)
Residential (2.78)

Preference was discussed in terms of posited preference predictors, Complexity, Coherence, Mystery, and Legibility, as well as content predictors of Nature and Water.

Manicured Landscapes had parklike natural scenes, high in ordered Complexity and with variations in texture and spaciousness that enhanced Coherence and Legibility.
Vegetation included natural scenes and Mystery but lacked the orderliness of Manicured Landscapes.
Pastoral consisted also of natural settings but lacked Complexity and Coherence, due primarily to the predominance of clear open spaces (spaciousness).
Residential scenes were characterized by Complexity but were rural (single-family houses), which, as in earlier studies, tended to lower preference.

Relation to "Expert" Predictors. Although the categories based on respondents' preferences and predictors made by the model were statistically related, there were some inadequacies in terms of the model's ability to predict the content underlying the least (Residential) and the most preferred (Manicured Landscapes) categories. Herbert attributes this to not enough attention having been paid to Coherence and Complexity predictor variables when establishing the land-use dimensions.

Noteworthy Points

Effort to use people's preferences in expert decision making by analyzing preferred areas for underlying content and inherent predictors of preference that can then be used to guide the location, preservation, or more sensitive planning of similar scenes

Effort to relate a physical model to visual preference, in order to predict and evaluate visual resource quality

A COGNITIVE ANALYSIS OF PREFERENCE FOR FIELD-AND-FOREST ENVIRONMENTS

Thomas R. Herzog (1984)
Landscape Research, 9(1), 10–16

Overview. The purposes of this study were (1) to identify kinds of environments or content categories that underly preference and (2) to investigate predictor variables that help account for preference in natural environments. Viewing time was also examined as a variable affecting preference. The researcher discusses the implications of these findings for planners incorporating nature in urban settings.

Photographs. The 100 color slides used were representative of field-and-forest settings in lower Michigan. They depicted relatively flat terrain and generous amounts of foliage. Photographs did not include people or water.

Sample/Survey. Participants were 247 undergraduate introductory psychology students. Each participant rated the slides on one of six predictor variables – Identifiability, Coherence, Spaciousness, Complexity, Mystery, Texture – or on preference. Sample size for each predictor variable was 21 or 22. Preference was rated with either 15 sec of viewing (N = 75) or one of two brief durations, 20 or 200 msec (N = 20 and 21).

Definition of Predictor Variables

Coherence - how well scene "hangs together"; how easy to predict from one portion of scene to another
Complexity - extent to which scene presents information that promises more information if only one had more time to look at it from present vantage point
Identifiability - sense of familiarity (rather than actual familiarity); how easy to get to know the scene
Mystery - extent to which scene promises more information if viewer could walk deeper into it
Spaciousness - sense of space; how much room to wander
Texture - how fine-grained ground surface of surface or obstruction is

Results. Using SSA-III (and criterion of .40 loading on any given category, no loading greater than .35 on another category), three categories were found: Unconcealed Vantage Point consisted of open spaces; Concealed Vantage Point consisted of views into open areas from within the forest; Large Trees consisted of views where large trees predominated, often seen from within the forest or bordering paths. Table B.3 summarizes the mean ratings for each category on each of the predictor variables as well as on preference.

Table B.3. *Category means and predictor ratings (Herzog, 1984)*

	Unconcealed Vantage Pt.	Concealed Vantage Pt.	Large Trees
No. of scenes	36	13	10
Preference (15 sec)	3.32	3.47	3.95
Identifiability	3.32 (.55)[a]	2.98	3.79 (.79)
Coherence	2.96 (.47)	2.39	3.98
Spaciousness	3.10 (.45)	2.21	3.24 (−.69)
Complexity	3.23	3.59	3.04
Mystery	2.76	3.19	3.57 (.91)
Texture	2.64	2.36	3.69

[a]Significant correlation between predictor and preference.

Viewing Time. The pattern of overall preference means did not change as a function of viewing time. The average preference rating for the two shorter duration times, however, was significantly lower than the mean for the 15-sec condition.

Noteworthy Points

Suggestions for planners in terms of design and preservation:
- preference for large, old trees
- positive influence of pathways, especially in combination with large trees as border elements

Mystery as important preference predictor

A COGNITIVE ANALYSIS OF PREFERENCE FOR WATERSCAPES

Thomas R. Herzog (1985)
Journal of Environmental Psychology, 5, 225–241

Overview. This study focused on preference for a variety of waterscapes. Analysis of preference for scenes indicated that the type of waterscape, viewing time, and six preference predictors as identified by an informational approach to preference – Texture, Coherence, Complexity, Spaciousness, Mystery, and Identifiability – all play a role in determining preference. In the most preferred grouping, characterized by rough surface textures, Mystery, Coherence, and Spaciousness positively predicted preference.

Site Photographs. The broad sample of 70 color slides, ranging from waterfalls and mountain streams to swamps and stagnant creeks, came from all over the United States. The photographs included no people, and human influences were minimal.

Sample/Survey. Participants, all psychology students, were divided into groups that rated the scenes on one of the predictor variables and groups that rated the scenes in terms of preference. Among the preference-rating groups, four had viewing times of 15 sec, whereas two others saw the slides for either 20 sec or 200 msec. All ratings used 5-point scales. The slides were shown as two sets of 40 slides with a break halfway. Five of the slides were repeated (and not analyzed).

Definition of Predictor Variables

Coherence - how much scene "hangs together"; how easy to predict from one portion of scene to another
Complexity - extent to which scene presents information that promises more information if only one had more time to look at it from present vantage point
Identifiability - sense of familiarity (rather than actual familiarity); how easy to get to know the scene
Mystery - extent to which scene promises more information if viewer could walk deeper into it
Spaciousness - sense of space; how much room to wander
Texture - how fine-grained ground surface is (or, if the ground surface is obscured by objects in the foreground, then how fine-grained surface of obstruction is)

Results. Split-half reliability measures were .91–.99 for preference (depending on viewing time) and .79–.92 for the predictor variables.

Categories. Categories formed by SSA-III analysis (factor loading $>.40$ on one category, $<.35$ on any other) were Mountain Waterscapes; Swampy Areas; Rivers, Lakes, Ponds; and Large Bodies of Water (validity was checked by having 18 psychology students group 57 of the scenes into categories based on verbal descriptions: rates of .83, .92, .80, and .97, respectively). Table B.4 summarizes mean ratings for each predictor variable and preference for each of these categories.

Table B.4. *Category means and predictor ratings (Herzog, 1985)*

	Mountain	Large Bodies	Rivers	Swampy
Preference (15 sec)	3.99	3.28	3.11	2.13
Spaciousness	3.11	4.11	2.95	2.45
Texture	2.05	3.80	3.20	2.69
Coherence	3.38	3.66	3.20	3.07
Complexity	3.39	2.08	2.87	3.44
Mystery	3.25	2.42	3.23	3.24
Identifiability	2.43	3.22	3.40	2.64

Predictors. Multiple regression analysis showed that the variables accounted for almost half the variance, with Spaciousness, Coherence, Mystery, and Texture making significant contributions. In general, scenes high in the first three but low in Texture were more preferred.

Analysis of the relative contributions of the categories and predictor variables included the 57 scenes falling into the categories. Together, the predictors and categories accounted for 86% of the preference variance. Only Coherence and Complexity made significant contributions. Several regression analyses established that the content categories are more potent predictors than the rated variables.

Viewing Time. As a main effect, viewing time was not significant. The significant interaction with the perceptual categories, however, was due to the greater preference for the least preferred Swampy Areas at shorter durations. (For the most preferred Mountain Waterscapes, the longest and shortest durations showed no preference difference, with 20 msec significantly lower.)

Noteworthy Points. Because waterscapes are generally preferred, they are often omitted from studies of natural settings. This study focused on the variety of waterscapes per se.

Introduction of concept of "tranquility" as a mediating quality between spaciousness and preference
Most thorough statistical analysis of preference predictors

A COGNITIVE ANALYSIS OF PREFERENCE FOR NATURAL ENVIRONMENTS: MOUNTAINS, CANYONS, AND DESERTS

Thomas R. Herzog (1987)
Landscape Journal, 6, 140–152

Overview. This study had two objectives: (1) to examine preferences for settings with uneven terrain; and (2) to test the general usefulness of the Kaplan information model in such settings. Preference for uneven terrain, defined here as mountains, canyons, and desert rock formations, was studied as a function of content categories, viewing time, and six predictor variables. The overall results uphold the informational model and suggest a number of directions for further research.

Photographs. The 70 color slides sampled environments from three categories: mountains, canyons, and deserts. They included no people and minimal evidence of human influence. The settings were drawn from southwestern and western states.

Sample/Survey. The sample consisted of 274 introductory psychology students at the University of North Carolina. They were divided into 12 groups of 13–26 people. Participants rated the slides, using a 5-point scale, for one of the seven variables. Sample sizes for the six predictor variables (defined below) were 22–24. For the preference ratings, 85 participants were shown the scenes at 15 sec. Shorter durations (20 and 200 msec) were also used for preference ratings only (N = 28 and 25).

The first nine slides were used for orientation. Slides were then presented in two sets of 40 (with the first three and last two in each set used as fillers). Four viewing orders were used.

Definition of Predictor Variables

Coherence – how well scene "hangs together"; how easy it is to get to know the scene

Complexity – scene contains many elements, promises further information if only there were more time to look at it from present vantage point

Identifiability – sense of familiarity (rather than actual familiarity); how easy it is to get to know the scene

Mystery – promises further information if one could walk deeper into scene

Spaciousness – feeling of space; how much room to wander around

Texture – how fine-grained ground surface or surface of obstruction is

Results. SSA-III produced five categories. For 15 sec viewing time, all but Narrow Canyons were equally preferred. Table B.5 shows the mean ratings for each category. Herzog examined the role of the predictor variables in three stages:

1. category differences in the predictor variables (using MANOVA)

Table B.5. *Category means and predictor ratings (Herzog, 1987)*

	Snowy Mtns.	Smaller Mtns.	Spacious Canyons	Desert	Narrow Canyons
No. scenes	15	4	4	20	8
Preference	3.31	3.31	3.38	3.26	2.94
Coherence	3.57	3.03	3.72	3.20	3.17
Complexity	2.43	3.01	3.62	3.18	3.26
Identifiability	2.94	3.61	2.57	2.66	2.39
Mystery	3.12	3.41	3.27	3.28	4.00
Spaciousness	3.53	3.51	3.79	2.66	1.96
Texture	3.10	2.53	2.81	2.56	2.36

2. role of the predictors in accounting for preference *across* the categories (using regression analysis)
3. role of the predictors *within* categories (using correlations)

Using all 70 scenes in the regression analysis, the predictor variables accounted for 60% of the variance in preference ($p < .001$), with significant contributions by Spaciousness, Mystery, and Identifiability. In general, scenes high in these predictors were more preferred. In ensuing analyses using only those scenes in the content categories, Identifiability remained a strong positive predictor of preference; Spaciousness and Mystery no longer predicted preference; Complexity emerged as a significant positive predictor, and content categories as significant predictors. (These analyses all used unweighted preference means, thus incorporating the shorter viewing times.)

A striking finding of the within-category analyses was the negative influence of Mystery on preference in the Narrow Canyon Category. Herzog suggests that for certain kinds of environments, when Mystery is part of a "package" that suggests immediate danger (i.e., high Mystery, low Spaciousness, very close vantage point, and knowledge of what might happen in such a setting), the typical positive effect of Mystery is overriden.

Viewing Times. There was no main effect of viewing time on preference. However, viewing time and preference showed a significant interaction, with preference for Snowy Mountains lower at 15 sec than for the shorter times; for Narrow Canyons the 20 msec preference was lower than the longer viewing times. The author explains these in terms of the informational model: Upon further examination, a "postcard" mountain view loses some of its appeal whereas having time to further examine a narrow canyon yields some "reassurance." The overall pattern of results supports the conclusion that reliable preference judgments and preference differences can be obtained from very brief viewing times.

Noteworthy Points

Clear explanation of extensive statistical analysis of data and insightful discussion of the implications and complexity of that analysis
Influence of viewing time on preference ratings
Evidence for Mystery as negative factor in preference for certain kinds of environments
Problematic efforts to use "texture" as a construct for ease of "locomotion"
Herzog mentioned the need for future research regarding perceived security/danger as variables in preference and further examination of environments (both urban and natural) where Mystery has a negative impact on preference

Ed. note. Herzog & Smith (1988) is such a follow-up study with interesting findings.

A COGNITIVE ANALYSIS OF PREFERENCE FOR URBAN NATURE

Thomas R. Herzog (1989)
Journal of Environmental Psychology, 9

Overview. This study had two objectives: (1) to examine preferences for urban environments with substantial natural elements; and (2) to test the general usefulness of the Kaplan information model in such settings. The overall preference for nature content was evident from the higher preference for the nature-related perceptual categories. Of the informational predictors, Coherence and Mystery were highly significant.

Photographs. The 70 color slides sampled urban environments reflecting a variety of natural elements, varying in their state of maintenance. No settings contained water or people. A range of building ages was included. Most scenes were taken in and around Grand Rapids, Michigan.

Sample/Survey. The sample consisted of 354 undergraduate students at Grand Valley State University. Twenty-seven sessions were conducted with groups of 5–21 participants. Participants rated the slides, using a 5-point scale, for one of 10 variables. Sample sizes for the predictor variables (defined below) were 25–29. For the preference ratings, 76 participants were shown the scenes at 15 sec. A 100 msec viewing time was also included for preference ratings only ($N = 34$). The first five slides were used for orientation. Slides were then presented in two sets of 40 (with the first three and last two in each set used as fillers). Three viewing orders were used.

Definition of Predictor Variables

Age – how old the elements in the scene seem to be
Coherence – how well the scene "hangs together"; how easy it is to organize and structure the scene
Complexity – how much is going on in the scene; how much there is to look at; extent to which scene contains many elements of different kinds
Legibility – how easy it would be to find one's way around in the environment depicted, to figure out where one is at any given moment or to find one's way back to any given point in the environment
Mystery – promises more to be seen if one could walk deeper into scene
Nature – how much foliage or vegetation there is in the scene
Spaciousness – feeling of space or depth the scene conveys; how much room there is to wander; extent to which the structure of the scene suggests that one would have to go a long way to reach its farthest point
Typicality – extent to which the scene seems to be a representative example of its class; how good an example the scene is of whatever category it belongs to

Results. SSA-III produced four categories. (Criterion for inclusion consisted of factor loading of at least .40 on one category only.) For 15 sec viewing time, the Contemporary Buildings and Concealed Foreground categories were equivalent in preference; other differences are significant. Table B.6 provides summaries of ratings for each category. Differences between the perceptual categories were significant with respect to each of the predictor variables, except for Typicality. As is evident from the equivalent ratings on Nature, both Tended Nature and Concealed Foreground were high in nature content. In the latter case, the concealment was achieved by the vegetation.

Table B.6. *Category means and predictor ratings (Herzog, 1989)*

	Older Buildings	Contemp. Buildings	Concealed Foreground	Tended Nature
No. scenes	12	7	10	8
Preference (15 sec)	1.53	2.64	2.83	3.58
Age	4.10	2.60	3.21	2.90
Coherence	1.84	3.23	2.82	3.63
Complexity	2.32	3.00	2.66	2.83
Legibility	1.99	3.30	2.64	3.30
Mystery	2.89	2.67	3.40	3.31
Nature	3.22	2.59	3.79	3.74
Refuge	3.58	2.26	2.97	2.54
Spaciousness	2.03	2.61	2.45	3.05
Typicality	3.17	3.29	3.01	3.20

Using all 70 scenes in the regression analysis, the predictor variables accounted for 87% of the variance in preference ($p < .001$), with significant contributions by Coherence, Mystery, and Nature. The same pattern held when only scenes in the four perceptual categories were included in the regression analysis.

Viewing Time. There was a significant preference advantage for the shorter viewing time, with a significant interaction with environmental category; Older Buildings and Concealed Foreground scenes were particularly preferred at the short durations. The overall pattern of preferences, however, is not affected by duration.

Noteworthy Points

Data analysis involves insightful and clear explanation of statistical procedures, including (as is true in earlier studies) the reliability of measurement for the predictor variables and use of analyses "by scenes" as well as in terms of participants, as appropriate.

Age of buildings was again an important element in differentiating among the categories, with Older Buildings once again least preferred. Whether this effect is related to the particular buildings in the study or to the participant sample requires further study. Scenes highest in nature content were highest in preference.

Typicality, here used as a substitute for "sense of familiarity," was the least "useful " predictor.

The very high correlations between Coherence and preference are surprising.

APPENDIX B THE PREDICTION OF PREFERENCE FOR FAMILIAR URBAN PLACES

Thomas R. Herzog, S. Kaplan, & R. Kaplan (1976)
Environment and Behavior, 8, 627–645

Overview. This study looked at how people perceive and respond to familiar urban places – their preferences, categorization of places, and the role of complexity and familiarity in accounting for those preferences. Results show that the experience of the urban environment is not unidimensional: The basis for place categorization does not depend solely on form or on function; the impact of Complexity and Familiarity is not uniform but must be viewed within the context of other factors.

Site Photographs/Sample. Eighty-six color slides of urban scenes of the Grand Rapids, Michigan, area were selected. Criteria included that the site have an identifiable object or structure that was relatively familiar to participants and identifiable by name and/or location. Functional categories included civic/government buildings, educational buildings, religious centers, retail outlets, business/office structures, theaters, hospitals, factories, and restaurants. Participants were 121 introductory psychology students who were relatively familiar with the sites.

Survey. Participants rated each scene on a 5-point scale in terms of Familiarity, preference, and Complexity. There were three methods of scene presentation: Slides (slide with name and location; $N = 74$), Label (name and location only; $N = 22$), and Imagery (label with time to imagine the place; $N = 25$). Two presentation orders were used. Time between each slide/label was about 15–20 sec.

Definitions of Predictor Variables

Complexity – intricateness, or the opposite of simplicity
Familiarity – how well known the scene is; personal experience or secondhand information from TV, and so forth

Results

Categories. Because over 10% of the sample did not recognize them, 16 scenes were not used in the analysis. Five categories emerged from analysis, using SSA-III, of the slide preference ratings of the remaining 70 (factor loading >.40 and <.35 on any other category). The categories and mean preference ratings were: Campus (3.6); Contemporary (3.2); Older Commercial/Service (3.1); Entertainment (2.5); Cultural (2.3). Visual preferences for the five categories were significantly different (F = 23.24, df = 4, 92, p <.001). Familiarity and Complexity also varied significantly for these categories (F = 26.50 and F = 9.21; p <.001, respectively).

Correlational analysis of Familiarity and Complexity as predictors of preference also differed for the content realms. Both Familiarity and Complexity correlated with preference (.51 and .43, respectively) and together accounted for 48% of the preference variance. This pattern was not characteristic within the categories themselves. For example, preference and Familiarity were negatively correlated with the Campus and Contemporary categories and positively correlated with the others.

Conditions. There were no significant differences in the ratings for the various conditions. Pairwise correlations for preference were in the low .80s; for Familiarity, ranging from .81 to .86; and for Complexity, between the nonvisual (Label and Imagery) and slide conditions, .64 and .66.

Noteworthy Points

Cautions against using global predictions of preference, considering instead the importance of content domains
Familiarity as variable in study of urban visual experience and component in establishing "sense of place"
For familiar places, responses to photographs and to the names of the pictured places are equivalent

THE PREDICTION OF PREFERENCE FOR UNFAMILIAR URBAN PLACES

Thomas R. Herzog, S. Kaplan, and R. Kaplan (1982)
Population and Environment, 5, 43–59

Overview. Continuing earlier research on the categorization of familiar urban places, this study explored preference in terms of how people categorize unfamiliar settings, using four predictor variables, Complexity, Coherence, Identifiability, and Mystery. An additional methodological question of viewing time was incorporated into the study design. Areas low in Coherence and high in Complexity are not preferred; and Mystery, as highlighted by an unexpected dimension, Urban Nature, enhances preference.

Site/Photographs/Sample. The 140 color slides depicted urban scenes throughout lower Michigan. They represented diverse urban settings but did not include scenes with people. Participants were 249 introductory psychology students. They were divided into 15 groups of 9–26 people.

Survey. The study consisted of two phases. Phase 1 participants indicated any scenes they recognized. Those who knew more than 10% of the scenes and those scenes recognized by over 10% of the sample were eliminated from the study. Students also rated (on a 5-point scale) each of the 140 color slides on two of the four predictor variables: Identifiability, Coherence, Complexity, and Mystery. Each of the six groups responded to a different combination of variables, so that each variable was rated by three groups.

Definition of Predictor Variables

Coherence – extent to which scene "hangs together" or contains repeated elements, textures, and structural factors that allow one to predict from one portion of a scene to another

Complexity – extent to which scene contains many elements, regardless of their arrangement; further information promised if time available to view scene from present vantage point

Identifiability – how easily one can tell what is being depicted; how easy it is to come up with an appropriate verbal description or label for the scene

Mystery – extent to which scene promises further information based on a change in vantage point of the observer such that observer is permitted to walk deeper into the scene

Phase 2. The mean ratings for each predictor variable for each photograph from Phase 1 were computed and ranked. The 70 slides were chosen from this ranking in order to sample the full range of ratings for the predictor variables. Phase 2 participants rated (5-point scale) these 70 slides according to preference, how much they liked each scene.

Viewing time. Four experimental conditions in Phase 2 were based on viewing times of 15 sec, 20 msec, two consecutive 10 msec viewings, and 10 msec.

Results

Categories. Using the 15 sec viewing time in Phase 2, SSA-III analysis yielded five categories (factor loading >.40 and <.40 on any other category). Table B.7 presents the summary information for each category.

Preferences were significantly different ($F = 265.92$, $df = 4$ and 564, $p < .001$). The relationship of preference to the predictor variables was examined both across dimensions and within them (correlations significant at $p < .05$). Observations included low preference for Alley/Factory, which rated high in Complexity and low in Coherence. Within this dimension, however, Coherence was a positive predictor of preference ($r = .61$). That Mystery was high in the most preferred dimension, Urban Nature, suggests it may be a potentially important factor in enhancing the urban setting. In Unusual Architecture, Identifiability and preference were positively correlated ($r = .81$), suggesting that people want to be able to make sense of new elements in the environment.

Effects of Viewing Time. The pattern of preferences was the same for the briefer viewing times as for the 15 sec with respect to the perceptual groupings. For the short durations combined, however, preference was significantly higher than the mean in the longer duration across all slides ($t = 2.34$, $df = 141$, $p < .025$).

Table B.7. *Category means and predictor ratings (Herzog, Kaplan, & Kaplan, 1982)*

Category	Pref.	Ident.	Coh.	Compl.	Myst.
Urban Nature	3.56	2.44	2.48	3.03	4.02
Unusual Architecture	3.16	2.36	3.45	3.28	2.92
Older Buildings	3.14	3.14	3.56	3.05	2.51
Contemporary Life	2.66	3.30	3.44	3.01	2.65
Alley/Factory	1.64	2.95	2.41	3.68	3.28

Noteworthy Points

Importance of nature in urban setting
Preference as function of what a setting affords
Experience of urban settings not just related to function, as suggested by planner's emphasis on land use (implicit difference between expert and nonexpert perception)
Example of methodology to incorporate public perception
Age of buildings as factor in perception of the urban environment

VISUAL PREFERENCE AS A TOOL FOR CITIZEN PARTICIPATION: A CASE STUDY OF URBAN WATERFRONT REVITALIZATION IN BURLINGTON, VERMONT

Thomas R. Hudspeth (1982)
Doctoral dissertation, University of Michigan (Rachel Kaplan, chairperson)

Overview. This study examined citizen reactions to various urban waterfront views and compared them with the citizen preferences as predicted by professionals. Results suggest that using visual preference can be an effective tool for facilitating citizen participation. This specific case study offers an example of how to apply visual preference to a waterfront situation as well as a number of specific approaches to revitalizing the Burlington waterfront area.

Site/Photographs. Thirty-two black-and-white photographs of the most heavily urbanized portion of the Burlington waterfront and eight photographs from two other waterfronts (Savannah and Alexandria) constituted the visual portion of the survey. All pictures were taken at eye level and included no people.

Survey/Sample. The mail survey questionnaire included a visual portion asking for respondent's preference (scale 1–5) for each of the photographs. Verbal items included questions about people's use of the waterfront, their satisfaction and problems with it, changes they would like to see. The survey was mailed to Burlington residents, business interests, professionals, government officials, and citizen groups. The final sample consisted of 431 respondents. Nine influential persons, from business, government, and the professional sector, were personally interviewed and asked to predict how they thought participants would group similar waterfront scenes and how they would rate them in terms of preference.

Results

Preference. Mean preference ratings for scenes ranged from 1.36 to 4.54. The most preferred were from the other waterfront sites and thus depicted "possible futures" for the Burlington waterfront. All included park and recreation activities. The least preferred were of industrial sites, utilities, parking lots, etc.

SSA-III and ICLUST analytic procedures produced five categories: Industry (preference, 1.84), Boats (3.62), Undeveloped (3.66), Parks and Recreation (3.98), Historic Preservation/Commercial (4.13). These groupings differed both from traditional land-use categories and from the predictions made by the influential citizens. Although they correctly predicted that industry was not well liked, in general they had only moderate success in speaking for the public at large.

Examination of preferences by subgroups in terms of both demographic characteristics and the groups mentioned above revealed more similarities than differences. Cluster analysis was also used with the five sets of verbal questions relating to Activities, Satisfaction, Problems, Changes and Changes for Burlington Specifically. Although active use of the waterfront was low in general, frequencies varied for the different Activity clusters: Socialize, Relax, Passive Recreate (2.88); Eat Out, Shop (2.05); Go Fishing, Boating (2.60); and Watch Trains, Smell of the Grain Mill (1.51).

People were not very satisfied with the waterfront at present. The Satisfaction clusters were: Pedestrian Access (2.85), Auto Conveniences (2.63), and General Attractiveness (2.47).

Rating of Problems included concerns about the area being Insufficiently Park-Oriented (3.66), the presence of Negative Environmental Factors (3.27), and Lack of Safety (2.56).

Ratings for clusters dealing with potential Changes indicated that people saw as desirable: Beautification (4.35), incorporating People-Use Attractions (3.53) (i.e., future uses that would attract people, such as a museum, farmers' market, etc.), and Commercial, Residential, and Public Development (3.17).

Categories for suggested Changes for Burlington Specifically included reactions to Specific Projects (3.64), such as the waterfront revitalization, a new civic center, library addition; Outdoor Sales (3.50), such as a farmers' market; and Growth and Development (3.27), which included possibilities of increasing tourism, number of jobs, population, and industry in the area.

Noteworthy Points

Comparison of citizen reactions to experts' predictions of citizen preference

Extensive bibliography on waterfront revitalization, citizen participation, and the Vermont situation specifically

Although the similarities in preference suggest that there are some general preferences to guide waterfront revitalization, Hudspeth warns against the "universal" solution and outlines some ways for smaller cities to retain and emphasize the special character of their area. Preferences indicated the value people place on historic preservation and adaptive reuse, and on the importance of both visual and physical access to waterfront recreation.

Ed. note. Portions of the dissertation can be found in T. R. Hudspeth (1986), "Visual preference as a tool for facilitating citizen participation in urban waterfront revitalization," *Journal of Environmental Management, 23,* 373–385.

PREDICTORS OF ENVIRONMENTAL PREFERENCE: DESIGNERS AND "CLIENTS"

Rachel Kaplan (1973)
In W. F. E. Preiser (Ed.), *Environmental design research.* Stroudsburg, PA: Dowden, Hutchinson, & Ross

Overview. This study employed both graphics and photographs to examine Mystery and Coherence as predictors of landscape preference and to look at preference prediction for people with different levels of training in design-related professions. Both Coherence and Mystery emerged as relatively independent and as effective predictors of preference. The preference predictions especially between the "designer" and "client" groups were different enough that the author warns, "if designers and the public are to work together to some degree on design decisions, a mutual recognition of these differences is essential."

Site/Photographs/Visuals. A series of 60 slides included 30 graphic renditions and 30 monochromatic photographs of four content areas: paths and highways, nature, buildings, parts of buildings (private and public). An effort was made to take photographs that matched the graphics. Graphics conveying a strong sense of texture as well as sketchier line drawings were included. The slides were also categorized by organizational qualities, by whether or not one could "enter" the scene (deep/shallow), and how spacious or enclosed it seemed (open/closed). In general, combinations of these two qualities were represented by 9–16 slides.

Sample/Survey. The 107 participants were drawn from landscape architecture (30) and architecture (38) students, the designers, and college students, the clients (39). The slides were shown for 20 sec, during which time participants rated (scale of 5) for Mystery ("To what degree do you think you would learn more if you could walk deeper into the scene?"), Coherence ("To what degree does it hang together?"), and preference ("How much do you like it?").

Results

Categories. Three domains emerged using SSA-III and ICLUST analysis. No graphic renditions were included in the first two.

Nature – natural setting with little human influence; characterized primarily as "deep" and enclosed
Part Buildings with Nature – part building in a distinctly natural setting; open and both shallow and deep
Building Complexes – graphic renditions of architecturally striking groupings; shallow

Graphics/Photographs. The mean preference ratings of the graphics and photographic scenes were virtually identical. The perceptual categories, however, suggest that the graphic renditions are seen as distinctly different.

Group Differences: Preference. Comparisons among the groups were based on the separate ratings (Mystery, Coherence, preference) and for the preference dimensions (Nature, Part Buildings, Building Complexes). The patterns that emerged suggest strong differences in preference reflecting professional interest: buildings for architects, buildings and landscape for landscape architects, and more natural settings for the general college students.

Preference Predictors. Mystery and Coherence were weakly related to each other for each sample. Partialing out the effect of preference, these correlations were −.32 (landscape architects), −.42 (architects), and −.58 (college students). Each of these predictors was strongly related to preference, however. Here the partial correlations showed the heaviest emphasis on Coherence on the part of the architects (r's .89 for Coherence and .66 for Mystery), and on Mystery for the college students (r's .72 and .93, respectively). For the landscape architects, the two predictors showed less variability (r's .72 and .80, respectively).

Noteworthy Points

Coherence and Mystery as predictors of preference
Issue of expert versus client: differences in preference and what "qualities" of a scene are important in that evaluation
Potential problem of having both preference and predictors rated at the same time

SWIFT RUN DRAIN STUDY

Rachel Kaplan (1977)
"Preference and everyday nature: Method and application," in D. Stokols (Ed.), *Perspectives on environment and behavior: Theory, research, and applications.* New York: Plenum

Overview. The scenic aspects of storm drains, although they are ubiquitous urban waterways, have received little research attention. This study, commissioned as part of a project to analyze modifications and improvements to an Ann Arbor, Michigan, drain system, examined nearby residents' perceptions of the drain and other waterways. Among the findings was the realization that the various regions (and drain treatments) have differences in preferences and problems; that verbal and visual responses can elicit very different reactions (even for the same kind of drain treatment alternative). Nevertheless, some "universal" preference predictors emerged as well.

Site/Photographs. The 32 black-and-white photographs included scenes taken along the Swift Run Drain in Ann Arbor, Michigan, as well as in other locations. They represented various treatment alternatives; for example, the drain water piped underground, as a creek, or as an impoundment or pond. Some scenes also showed flooding or neglect of the area near the waterway.

Sample/Survey. Of the 246 residents living in communities along the drain who received a written questionnaire, 115 responded. The 10-page survey included six pages of verbal questions about people's use of and reactions to the drain and alternative treatments as well as background information. Participants were also asked to rate four pages of photographs (eight to a page), using a 5-point scale, on (1) how similar the pictured waterway was to the one near their home and (2) how much they would like the waterway near them to look like the one pictured.

Results

Categories. Content-Identifying Methodologies (SSA-III and ICLUST) produced four categories that suggest how people experience the drain: Covered Drain (preference, 2.77), no water; Creek in Parklike Setting (2.74), creek flowing through residential areas but houses less noticeable; Impoundment (2.27), large contained bodies of water; and Backyard Creek (1.74), creek again in residential setting but separated from residential property by fences and lack of maintenance.

Verbal/Visual Preferences. In general, preferences were highest for scenes with a sense of spaciousness and orderliness. There were several significant differences between people's responses to visual and to verbal analogues. One group of residents, for example, had a high preference for the picture but low preference for the statement of the Impoundment alternative. Similarly, "burying the drain" met with greater favor from residents in one community when presented as a verbal option than it did in response to photographs of covered drains.

Familiarity. Though those living near an impoundment and near the parklike covered drain favored such settings more than the group as a whole, the effect of familiarity on preference was not always positive. Having the least preferred Backyard Creek close at hand made such scenes no more likable.

Noteworthy Points

Role of spaciousness, sense of orderliness on preference
Familiarity has a complex relationship to preference
First use of photographic questionnaire
Importance of "finding out where people are at"; differences in verbal and visual interpretations and in terms of local preferences and problems

Ed. note. This study is also discussed in S. Kaplan & R. Kaplan (1989), "The visual environment: Public participation in design and planning," *Journal of Social Issues,* 45.

ALTERNATIVE STRATEGIES IN THE STUDY OF ROADSIDE PREFERENCE

Rachel Kaplan (1977)
"Preference and everyday nature: Method and application," in D. Stokols (Ed.), *Perspectives on environment and behavior: Theory, research, and applications.* New York: Plenum

Overview. This study evaluated how well several approaches to identifying scenic landscapes relate to what people see as "salient" (i.e., prefer) in a scene – land-form/landcover classifications, visual categorizations (identifying fore-, mid-, and background characteristics), and a statistical way of identifying visual content from people's expressed preferences. Because some of these methods are often determined by designers, the study also examined how well these a priori "merit" weightings compare to people's preferences. Results suggest that preference is more accurately predicted by assessing the visual content of the environment as people experience it.

Site/Photographs. The 80 photographs were of roads under consideration for a scenic highway designation in Michigan's Upper Peninsula. The photographs contained no water, hillsides, or industrial, commercial, or residential settings. The photographs selected satisfied criteria relating to:

land use – wild land, varied farmland/forest, continuous forest, and farmland
land form – bedrock, moraine, ground moraine, outwash plain/old lake bed
visual categorizations (as rated by judges) – Flat Farmland 1, Flat Farmland 2 (more scruffy), Midground Scruffy, Forest/Midground, Forest

The photographs were organized into 20 sets (four to a set) to permit comparison of different combinations of land-use, land-form, and visual categorizations.

Sample/Survey. The survey was administered to 61 people, both locals and nonlocals, in the Upper Peninsula. They were asked to rate (5-point scale) each of the 20 arrays on how much they liked the composite pictures.

Results

Background Variables. There were few differences in preference in terms of age or sex. The young tended to like the most preferred scenes more, and the elderly did not dislike the least preferred as much. Overall preferences were significantly lower for locals than for nonlocals ($t = 2.64$, $df = 58$, $p < .01$).

Categories. ICLUST and SSA-III produced three categories: Flat Open Farmland (2.74 mean preference), Dense Forest (3.40), and Open Forest (3.72). Each of these categories had representative photographs from several land-form, land-use, and/or visual categorizations. The results suggest that spaciousness is an important aspect of preference. The four most preferred scenes (4.55) similarly depicted relatively open woodland.

Internal Coherence of Categorizations. Measurements of the internal consistency of the land-use/land-form categorizations were "mediocre" (alpha .23–.63). The empirically derived dimensions were relatively high in coherence (between .76 and.84). Four of the five visual categories were also relatively high. (.70–.79).

Experts/People. The results support the expectation that forests would be preferred over Flat Open Farmland. However, in terms of specific land-use patterns, the preferences for certain forest types and land-form designations implicitly assumed in expert models were not found.

Noteworthy Points

Lower preference for locals than for nonlocals
Comparison of expert assumptions with use of photo album procedure

Ed. note. More complete discussion is in unpublished paper: R. Kaplan & S. Kaplan (1974), written in conjunction with K. J. Polakowski's project: "Upper Great Lakes Regional Recreation Planning Study, Part 5: Scenic highway system." Upper Great Lakes Regional Commission.

WILDERNESS PERCEPTION AND PSYCHOLOGICAL BENEFITS: AN ANALYSIS OF A CONTINUING PROGRAM

Rachel Kaplan (1984)
Leisure Sciences, 6, 271–290

Overview. Part of a decade of field research on a wilderness outing program, this study used a photo-questionnaire to supplement self-reports of the experience. Comparisons of preference and familiarity ratings of the photographs from before and after the experience suggest the pervasiveness of the wilderness experience. The richness of the psychological benefits of the wilderness consistently found in this research suggests that psychological dimensions also may be vital aspects of effective human functioning in other settings.

Photographs. Twenty-one of the 24 black-and-white photographs depicted the forest environment of Michigan's Upper Peninsula that participants encountered during the wilderness experience. The three "control" scenes represented typical Michigan roadside scenery.

Sample/Survey. The 49 participants included 22 adults (17 female, 5 male) and 27 high school students (15 female, 12 male). Each had been on one of the 9-day wilderness outings, held over a 2-year time span. Participants were asked to keep personal journals about their experience and to fill out questionnaires before and after the wilderness outing and immediately following a 1-day "solo." The questionnaires administered before and after the experience included 25 items reflecting moods and feelings as well as a photographic survey asking for preference and familiarity ratings on a 5-point scale. The questionnaire completed at the beginning of the outing also included items about participants' previous outdoor experience and the Environmental Preference Questionnaire (EPQ).

Results

Familiarity. For all but the three control scenes, familiarity ratings were higher in the postsurvey ($p < .001$). Significantly higher familiarity ratings of participants with more out-of-doors experience occurred only in the initial questionnaire. The higher familiarity ratings in the final survey of those who were high on the EPQ *Nature* cluster suggest that having a preference for a setting heightens one's sensitivity to it.

Preference. SSA-III analysis of the initial preference ratings of the photographs produced three groups: Open Forest (6 scenes, mean preference 4.26); Forest Edge (5, 3.77) and Swamp (4, 3.56). For the first two of those categories, preference at trip's end was no different than at the start. Those with more out-of-doors experience had higher overall preference ratings for the Open Forest and Forest Edge scenes. Again, the EPQ Nature rating was a good predictor for preference of these two clusters. The Swamp category, by contrast, was significantly less preferred (mean 3.07) after the wilderness experience. Although participation in the outdoor experience heightened participants' sensitivity to the environment (as suggested by increased familiarity ratings), the author stresses that this increased familiarity does not necessarily mean increased preference. Increased exposure can lead to decreased preference if the area in not preferred in the first place.

Noteworthy Points

Role of intensive outdoor experience on preference and familiarity
Exploration of relationship between preference and familiarity

NATURE AT THE DOORSTEP: RESIDENTIAL SATISFACTION AND THE NEARBY ENVIRONMENT

Rachel Kaplan (1985)
Journal of Architectural and Planning Research, 2, 115–127

Overview. This study surveyed residents of nine multiple-family housing complexes to identify perceptions of the kinds of nature in the urban environment; to see how important those areas are to people; and to see how preference for these areas relates to residential satisfaction. An underlying question was whether there are aspects of site arrangement in the multiple-family context that can achieve some of the benefits of the single-family home. Results suggest that the relationship between natural elements and housing is central to the perception of those places: Lack of buildings makes a setting more preferred; building-dominated areas are less preferred. Furthermore, sensitive landscaping can enhance preference.

Sites/Photographs. The nine multi-family, cluster-type housing complexes were selected to represent examples of small, medium, and large complexes with low and relatively high densities. All were two to three stories tall, built between 1964 and 1975. Three were cooperatives. The 40 photographs used depicted different aspects of nearby natural elements: open spaces, parks, natural areas.

Sample/Survey. Of the 810 photo-questionnaires distributed, 268 usable surveys were returned. Efforts were made to distribute surveys randomly within a stratification of those households facing natural areas and those facing nonnatural ones (e.g., street or parking lot). Most of the respondents were in their twenties, without children, and transient. Participants rated eight pages of photos (eight to a page) and five pages of verbal items. They rated the first five pages of photographs on two scales: (1) the availability of areas such as depicted (3-point scale); and (2) how much they liked the setting (5-point scale). The other three pages involved rating (5-point scale) in terms of how much one would like to do a chosen activity in that area. Verbal items explored the availability and importance of certain natural elements - for example, the view from the window - as well as neighborhood satisfaction, perception of the social mix of the neighborhood, and residential satisfaction in general.

Results

Preference for Kinds of Nature. SSA and ICLUST analyses produced six categories based on preference:

Parkland (pref = 4.2) - characterized by smooth ground textures
Nature (4.0) - a diverse group of natural areas found at edge of the complexes and in parks or designated nature areas
Landscaped (3.2) - plantings along buildings and walkways
Open, Residential (3.0) - larger areas with lawns, buildings, and some trees
Building-Dominated, Mowed (2.2) - neatly mowed, large expanses
Barren, Anonymous (2.0) - smaller, faceless areas

Noteworthy Points

Importance of nonactive or less active activities (gardening, view from the window) in preference for natural, open space
Role of the balance between built and natural, as well as the organization of space in determining preference

Ed. note. More detailed presentation of this material is in R. Kaplan (1981), *Nearby nature and satisfaction with multiple-family neighborhoods*, Report to the Planning Department, City of Ann Arbor, MI (35 pp.). Portions of this study are also included in R. Kaplan (1983), "The role of nature in the urban context," in I. Altman & J. F. Wohlwill (Eds.), *Behavior and the natural environment* (New York: Plenum). Discussion here focuses only on photo-questionnaire aspect of the study. Appendix D includes other aspects.

CULTURAL AND SUB-CULTURAL COMPARISONS IN PREFERENCES FOR NATURAL SETTINGS

Rachel Kaplan and Eugene J. Herbert (1987)
Landscape and Urban Planning, 14, 281–293

Overview. Preference and familiarity are examined in both a cross-cultural (American and Australian students) and subcultural (Australian students and conservationists) context. In addition to correlational analysis, the study looks at actual preference levels and perceptual categories. Though preference ratings were similar, the perceptual categories of the student samples differed. Interestingly, the Australian students and conservation club members had distinctly different preference patterns, showing another aspect of familiarity: knowledge and expertise. The authors suggest that preference ratings could prove an effective means of identifying differences among interest groups.

Sites/Photographs. Sixty color slides included five Western Australian landscapes: escarpments, uplands, valleys, plantations, and eastern woodlands.

Sample/Survey. Three groups participated: Two of them, 120 Australian students and 74 members of the Wildflower Society, shared a cultural perspective and relative familiarity with the landscapes. The third group, 145 American students, represented a different culture but were of similar age and educational status to the Australian students. The student groups saw one of two randomized orders of slides, viewed each slide for 10 sec, and rated (on a 5-point scale) how much they liked the scene. The Wildflower group saw only one order of slides.

Results

Preference. There was high agreement across all samples with correlations of the mean ratings: .84 (between students); .81 (between Australians); .65 (Americans/Wildflower). The most preferred scenes were lusher, with a mixture of water and forest; the least preferred received a 1.5 rating from all three groups. There were differences: The Australians in general had overall higher preferences ratings than the Americans (3.28 to 3.11). In 28 of the scenes, the differences were significant, eight of them at least .50 higher.

The Wildflower and Australian student samples also differed. The Wildflower group had significantly higher preference ratings for 22 scenes (10 of these had at least a .50

difference). For 11 scenes generally depicting pines and other nonnative vegetation, the students had higher preference (five scenes had ratings differing by more than .50). This difference highlights the Wildflower Society's involvement (i.e., greater knowledge and concern for species types) in an ongoing dispute over the introduction of exotic species in Australia.

Perceptual Differences: Categories. Using the ICLUST and SSA-III, five categories emerged using the American student results and three based on the Australian ratings. The American groupings of a "foreign" landscape were more differentiated and depended more on perception (presence of trees and ground texture). The Australian-based groupings characterize the Western Australian landscape of eucalypt forests, open arid scrublands, and pastoral, grazed, spacious areas. Table B.8 provides the mean ratings for each sample, based on both sets of CIMs.

Noteworthy Points

Multidimensional quality of preference; differential role of familiarity, but also all groups' preference influenced by spatial organization and content properties (i.e., presence of trees, water) regardless of familiarity

Role of interest and expertise (knowledge and concern) in defining subcultures, which are reflected in preferences

Table B.8. *Category means for each sample (Kaplan & Herbert, 1987)*

	Preference			
	American	Australian	Wildflower	p
American groupings				
Open Woodland/Field	2.64	2.94	2.96	.0001
Rough-textured, Arid, Wooded	2.67	2.74	3.20	.0001
Open, Smooth Texture	2.90	2.95	2.86	NS
Vista (not heavily wooded)	3.57	3.34	3.13	.0001
Forest/Forest Vista	3.88	4.00	3.97	NS
Australian groupings				
Arid, Open, Coarse Texture	2.44	2.61	2.89	.0001
Manipulated, Open, Spacious	3.13	3.08	2.85	.005
Trees in Forest	3.28	3.57	3.71	.000

APPENDIX B

FAMILIARITY AND PREFERENCE: A CROSS-CULTURAL ANALYSIS

Rachel Kaplan and Eugene J. Herbert (1988)
In J. L. Nasar (Ed.), *Environmental aesthetics: Theory, research, and applications,* New York: Cambridge University Press

Overview. The relationship between familiarity and preference is explored within the cross-cultural context. Participants from Western Australia and the United States rated rural, "nature" scenes from the United States. Results suggest that whereas familiarity can increase preference for areas already preferred it can "breed contempt" for less preferred scenes. Overall preferences for the scenes between the groups were very similar. However, categorizations and preference for some sets of scenes differed. This suggests the importance of having research methodologies sensitive not only to individual scenes but also to the categories implicitly defined by the scenes.

Site/Photographs. The 55 color slides depicted everyday, rural scenes from Oakland County, Michigan. The pictures presented a diversity of land forms and land uses, with no close-ups or people.

Sample/Survey. The American sample consisted of 97 psychology students (study completed in 1981 by Herbert); the 122 Australians were similarly enrolled in an introductory psychology course. Participants looked at one of two random orders of slide presentation. They viewed each slide for 10 sec and rated it (on a 1–5 scale) on "How much do you like the scene?"

Results. Correlation of preference ratings across the two groups was .84, with Americans generally preferring the scenes (means 3.29 and 3.08). The Australians' most preferred scenes were included in the American high-preference group, and nine of the Americans' least preferred were also among the Australians'. Americans rated 18 of the scenes at least .40 higher.

Categories. SSA III was used to derive perceptual categories from the preference data. Four groupings based on the Americans' ratings were Residential, Vegetation, Pastoral, and Manicured Landscapes; SSA-III analysis on the Australian sample yielded similar groups. Although the similarities are striking, both in preference and in dimensional groups, the differences in the relative ratings of the Pastoral/Open category suggest the complexity of the familiarity/preference relationship. The overall higher American preference ratings suggest that familiarity enhances preference. However, this dimension, the least preferred for the Australians, included scenes similar to the landscape in the nearby Australian countryside, suggesting that the more familiar is a less preferred environment. Table B.9 shows the overlap between these categories and the mean preferences for both samples.

Noteworthy Points

Need to examine both perceptual categories and preference to appreciate cross-cultural variations

Striking similarities in preference despite significant variation in flora of the two countries

ENVIRONMENTAL PREFERENCE: A COMPARISON OF FOUR DOMAINS OF PREDICTORS

Rachel Kaplan, Stephen Kaplan, and Terry J. Brown (in press) *Environment and Behavior*

Overview. This study examines the effectiveness of variables from four different domains in predicting preference.

Table B.9. *Preference means for both sets of categories (Kaplan & Herbert, 1988)*

	Pref. Amer./Aust.	No. slides	Overlap
American, Manicured	3.9/3.8	4	
Australian, Picturebook	3.9/3.8	3	−2, +1
American, Vegetation	3.5/3.4	17	
Australian, Vegetation	3.5/3.4	14	−5, +2
American, Pastoral	3.1/2.6	15	
Australian, Open Rural	3.1/2.7	21	−2, +8
American, Residential	2.8/2.8	10	−1, 0
Australian, Residential	2.9/2.9	9	

Three of the domains - Landcover Types, Informational, and Perception-Based - had variables with independently significant predictive power. Surprisingly, the fourth domain, based on the traditional approach to scenic assessment of using physical elements within a scene, did not yield any significant predictors. The authors discuss their findings in the context of other preference research and raise a number of issues for further inquiry.

Site/Photographs. The 59 photographs used were selected from a set of 243 photos of Scio Township, Michigan, which depicted a variety of land covers and land forms and whose location and orientation had been mapped. None of the photographs included residential, industrial, or commercial scenes or any water.

Sample. The photos were rated by undergraduate psychology students at the University of Michigan (N=88) and Grand Valley State College (N=92) for preference, using a 5-point scale.

Predictor Variables. A total of 20 variables were examined. Two of the domains were physically based:

Physical dimensions - related to elements of land form and land cover: Slope/Relief, Edge Contrast, Spatial Diversity, Naturalism, Compatibility, Height Contrast, Variety

Land-cover types - related to patterns: Agriculture, Cut Grassland, Weedy Fields, Scrubland, Forests, Woodlawn

The other domains focused on informational qualities of the spatial arrangement within the photographs:

Informational - Coherence, Complexity, Legibility, Mystery

Perception-based - Openness, Smoothness, Ease of Locomotion

The 59 photographs were rated by a team of two for variables in the physical (5-point scale) and land-cover (on a binary scale) domains. This team also rated the scenes on Openness and Smoothness (5-point scale). A five-person team rated the scenes for the informational variables and Ease of Locomotion.

Results. Regression analyses were run using variables within the four domains as independent variables and mean preference rating as the dependent variable.

Physical Dimensions. Of the four domains, only this one had no significant predictors. The authors posit that this is due to: (1) lack of topographical variation in the photo sample; (2) the fact that most of the scenes included vegetation, leaving little to test for Compatibility and Naturalism.

Land-Cover Types. Overall, the six types provided a highly significant basis for preference prediction (R^2 = .47). The least preferred types, Weedy Fields, Scrub and Agriculture (ratings 2.8–3.0), had significantly negative impacts on preference (r = .−53, −.39, −.35, respectively). The more highly preferred variables did not contribute significantly to the regression equation.

Informational. This was the lowest of the domains with significant regression results (R^2 = .19); only Mystery (partial r = .31) was a significant predictor.

Perception-Based. This domain was the most powerful predictor of preference (R^2 = .62). All three variables were significant with Openness (partial r = −.72) and Smoothness (partial r = .57) the stronger. The negative effect of Openness implies that scenes that are not wide-open are more preferred.

All Predictors. Regression analysis of all 20 variables accounted for a substantial amount of variance (R^2 = .83). The five variables selected in stepwise regression accounted for 69% of the variance. These variables – Weedy Fields, Scrubland, Mystery, Openness, and Smoothness (partial r = −.43, −.35, .41, −.39, .61) – were also significant in the separate domain analyses. (Agriculture, although significant in the land-cover type regression, was not a significant predictor in the 20-variable analysis.)

Noteworthy Points

Indication of independent predictive power of several preference assessment domains

Effort to compare and evaluate a number of approaches to scenic assessment

Replication of results: Mystery as significant information variable and diminished role of Complexity in preference

Poor performance of often-used physical dimensions to predict preference
Suggests importance of using a broader sampling of predictors in studying preference
Need for future study of differential effectiveness of predictors depending on type of landscape/scenes

ETHNICITY AND PREFERENCE FOR NATURAL SETTINGS: A REVIEW AND RECENT FINDINGS

Rachel Kaplan and Janet F. Talbot (1988)
Landscape and Urban Planning, 15, 107–117

Overview. Although research has shown that residential uses of various ethnic groups differ, results on environmental preference have been, for the most part, similar. Three studies examining some of the discrepancies in preference between blacks and whites focused on three questions: (1) Are there ethnic differences in environmental perception? (2) Can the variables underlying these differences be identified? And (3) if differences exist, are they merely a reflection of differing preferences for built or natural environments? The studies found that although nature in the nearby environment is important universally there are substantial differences in the types of setting preferred. These suggest the importance of having a number of different design and management strategies for urban open space to reflect the concerns and meet the needs of the various segments of society.

Site/Photographs. The article is based on three studies, which used black-and-white photographs of a variety of outdoor areas ranging from manicured parks to unmanaged, "wild" areas. Study 1 consisted of 37 photographs. A subset of 26 of these scenes was used in Study 2. Study 3 used 15 settings, each represented by four photographs from Ann Arbor.

Survey/Sample. For all three studies, participants sorted the photographs into five piles according to how much they liked them. In Study 1, people explained what they liked and did not like about the highest and lowest ranked piles. A checklist of these environmental features was used in Study 2. The procedure of Study 3 was similar to the first study, but participants also first grouped the settings according to size. Sample sizes were:

Study 1–31 (21 whites, 10 blacks from Ann Arbor)
Study 2–97 (black residents from Detroit)
Study 3–53 (47 whites, 9 blacks from Ann Arbor)

Results

Preference. Both groups expressed similar preferences for scenic parks, tree-lined residential streets, scenes with a few distinctive trees, rivers, and a single-family backyard. Ratings between blacks and whites, however, were significantly different for 30% and 47% of the scenes in Studies 1 and 3, respectively. The preference correlations for Studies 1 and 2 (using the same 26 photographs) between Detroit and Ann Arbor blacks was $r = .77$ ($p < .05$); between the Ann Arbor whites and Detroit blacks, $r = -.51$ ($p < .05$). Whites had higher preferences for scenes with dense foliage and overgrown vegetation. Many such scenes were disliked by blacks. Blacks significantly preferred scenes with paved walks and those that included some built structures; these were disliked by whites. The verbal comments supported these preferences. Over half the Detroit sample mentioned neatness (89%), the presence of water or trees (71% and 95%, respectively), as well as the incorporation of built elements such as benches and walkways (89%) as aspects of preferred settings. These comments also suggest that though urban blacks value having trees and nature nearby their preferences reflect concerns about orderliness, safety, and visibility within an area.

Categories. CIM analysis of the data (ICLUST and SSA-III) from Study 2 resulted in three groupings: Local Parks and Walks (mean preference 4.02), residential street scenes and small, well-maintained parks; Woods and Water (3.10), larger natural areas with lake or water in the foreground; and Woods and Path (2.98), similar to Woods and Water but with a pathway or small opening through the scene. The preferences for these dimensions again reflect black preference for smooth ground texture and well-kept, open areas with built features.

Importance of Nature. Blacks and whites did not differ significantly on ratings of how important nature contact is to them and how important that contact is near where they live; 77% of the Detroit sample rated nature as very important.

Noteworthy Points

Preference for natural environments in a nearby urban environment universally high

Ethnic (racial) differences related to the configuration of natural environments, with safety/danger being an important consideration for urban blacks

Photographs as an effective means of distinguishing differences in preference for kinds of natural settings

Ed. note. Other publications dealing with these studies:

Kaplan, R., (1984c). Dominant and variant values in environmental preference. In A. S. Devlin & S. L. Taylor (Eds.), *Environmental preference and landscape management*. New London: Connecticut College.

Talbot, J. F., & Kaplan, R., (1984). Needs and fears: The response to trees and nature in the inner city. *Journal of Arboriculture, 10*, 222–228.

RATED PREFERENCE AND COMPLEXITY FOR NATURAL AND URBAN VISUAL MATERIAL

Stephen Kaplan, Rachel Kaplan, and John S. Wendt (1972)
Perception and Psychophysics, 12(14), 354–356

SOME METHODS AND STRATEGIES IN THE PREDICTION OF PREFERENCE

Rachel Kaplan (1975)
In E. H. Zube, R. O. Brush, and J. G. Fabos (Eds.), *Landscape assessment: Values, perceptions, and resources.* Stroudsburg, PA: Dowden, Hutchinson, & Ross

Overview. The studies described here used the same set of photographs. The first study looked at the relation between Complexity and preference and how content (e.g., nature or urban) affects preference. The second one used the groupings generated in the first to further examine "qualities" in a scene that might play a role in preference – Complexity, Coherence, and Mystery. Finally, the third study focused on the influence of different viewing times on preference. In all, a high preference for nature content emerged. Complexity did not predict the nature-over-urban preference, although within the nature and urban groupings it was positively correlated with preference. This suggests the utility of separating content domains (here the natural and built) in analyzing such relationships. In the second study, Mystery was a very strong positive predictor of preference, especially for nature settings; and the researchers raised the possibility that preference ratings might bias other judgments. Correlations of ratings across different viewing times, examined in the third study, were very high, but results suggest that, with brief exposure, affective reactions are heightened – the more preferred are better liked; the less preferred are down-rated even more.

Site/Photographs. Three area types characterized the 56 color slides of nonspectacular, local places: urban scenes taken in Detroit and Ypsilanti, Michigan; nature scenes mostly from the University of Michigan arboretum; and a variety of mostly residential scenes. Slides were selected to equally sample four levels on a continuum from "natural" through "predominantly natural" and "predominantly human influenced" to "urban." These were randomly distributed in the viewing sequences.

Study 1 (Kaplan, Kaplan, & Wendt)

Sample/Survey. Eighty-eight women college students, 25–30 per session, rated the slides (5-point scale) on (1) how intricate or complex each one was, (2) how much they liked each one, (3) how intriguing each was, and (4) whether they wanted to look at the picture longer (indicated with a check). During the break between slide sets, participants performed pencil-and-paper tasks. Viewing time for each slide was 20 sec. The last two questions were so highly correlated with preference that they were not discussed in the findings.

Results

Categories. Cluster analysis using SSA-III for both preference and Complexity ratings produced two categories, Urban and Nature. These were similar to those established a priori except that the Nature category included some scenes from the "predominantly nature" group. Nature was significantly more preferred ($t = 8.45$, $df = 34$, $p < .001$); and Urban was seen as more complex ($t = 3.38$, $df = 34$, $p < .01$). Complexity did not account for preference; for the 56 slides the correlation of preference and Complexity was .37. They were significantly correlated within each dimension (.69 for Nature; .78, Urban). The residential scenes did not form a coherent group.

Study 2: Independent Ratings and Other Variables (Kaplan)

Sample/Survey. This follow-up study used the same set of slides but viewed them not as a continuum but in terms of Nature (22 slides), Urban (12), and others (22). The focus here was on the methodological issue of the independence of ratings and on what other variables might play a role in predicting preference. Separate groups of participants rated the slides (5-point scale) on one variable each: Mystery, Coherence, Complexity, and preference..

Definition of Predictor Variables

Complexity – how intricate the scene is
Mystery – promise of further information based on a change in vantage point of the observer; whether one would learn more if one could walk deeper into the scene
Coherence – extent to which the scene "hangs together"; repeated elements, texture, and structural factors

Results

Independence of Ratings. The Complexity ratings done in conjunction with preference (the first study) were compared with those done independently. The correlation of the two sets of Complexity ratings was $r = .79$; of the preference ratings, $r = .69$. Correlation of the preference ratings to the independently derived Complexity ratings showed a relatively strong negative correlation (−.52 and −.58) for the Nature and Urban slides combined, and no significant correlation within the categories. This contrast from the results in the first study raised the possibility that preference ratings biased the Complexity judgment.

Other Variables. As in the earlier study, preference for the Nature domain over the Urban was significant (t = 4.92 and t = 6.76, df = 34, $p < .001$), whereas Urban was seen as significantly more complex (t = 7.08 and t = 3.80, df = 34, $p < .001$). Coherence did not prove an important predictor. Mystery was a strong predictor of preference and was significantly higher for the Nature category (t = 4.09, df = 34, $p < .001$). The correlation of Mystery and preference for Nature and Urban combined was .64 (.53 and .55 within each, respectively). All three predictors combined led to R = .79 (.64 and .55 within each domain).

Study 3: Viewing Time (Kaplan)

Sample/Survey. Participants rated slides on preference only, with different groups viewing the slides for 10, 40, and 200 msec.

Results. There were no significant differences, with correlations of .97 between pairs of duration conditions. The preference predictors (using independent ratings of the slides from Study 2) approximated that predicted by the longer exposures of the earlier studies except for Mystery. Before, Mystery had positively predicted preference for both the Nature and Urban categories. With shorter durations, correlations were .41 and −.52, respectively. The shorter exposure appears to heighten response. The preference differences were more pronounced for Nature (F = 6.03, df = 2.53, $p < .005$) and for Urban (F = 4.86, df = 2.53, $p < .05$).

Noteworthy Points

Marks early identification of the preference predictors that underly much of the later "informational"-related preference research

Introduces informational attributes as predictors of preference

APPENDIX B RIVERSIDE PREFERENCE: ON-SITE AND PHOTOGRAPHIC REACTIONS

Judith Levin (1977)
Master's thesis, University of Michigan (Rachel Kaplan, chairperson)

Overview. An informational theoretical approach was used to explore preference for the "everyday riverside" view. Assessments of attractiveness were elicited both on-site and from photographs. Meaningful groupings of the scenes could be explained in terms of the four informational variables; those relating to understanding the environment - Legibility and Spaciousness - and to being involved there - Complexity and Mystery. The interplay of these variables also proved successful in predicting preference; settings rated higher for both the "making sense" and the "involvement" factors were more preferred.

Site/Photographs. Fourteen sites of the Huron River in and near Ann Arbor, Michigan, were selected to represent a diversity of habitats. The views were primarily natural, with limited evidence of human impact. To insure a common point of view, on-site participants looked through a 4 × 5 viewing camera, and all color photographs were taken from the river's edge.

Sample/Survey

On-site. The 38 participants (11 women, 27 men) were predominantly undergraduates interested in natural resources. They looked at each selected area and indicated their preference for the view, using a 5-point scale. In addition, participants described aspects of the scene that seemed especially striking and provided a +/− designation to each factor they mentioned. Background information included age, sex, and education.

Photographs. Twenty-four different students (primarily graduate, 8 women, 16 men) were asked to sort color photographs of 13 of the same river views into groups that seemed similar, to explain why, and to give their preference for, each grouping they proposed.

Results

On-site: Preference. Mean preference ratings for the riverscapes ranged from 2.66 to 4.13, most falling between 3.50 and 3.82.

On-site: Categories. Using ICLUST and SSA-III produced three categories (alpha .63–.87): Vista, view of a hill or bend in the river; view Obstructed by overgrown vegetation and narrow river width; and Lakelike, with bands of water, land, and sky. Levin's ratings of informational factors for the categories and mean preferences for each are shown in Table B.10.

Photographs. Four groupings of sites with the highest frequency of association were identified. All but one of them closely matched the groupings based on the on-site preferences. For these three, the mean preferences (Table B.10) are in the same rank order. Obstructed and Lakelike views, however, are much less preferred based on the photographs. The fourth photo grouping consisted of only two scenes, with strong vertical tree trunks contrasting to the water. These received very high preference ratings (mean 3.84). Again, informational variables proved useful in differentiating the groups and in predicting preference. The new, highly preferred cluster was high in both Complexity and Mystery and middling in the other two predictor variables.

Table B.10. *Category means and predictor ratings (Levin, 1977)*

			Involvement		Making sense	
Category	On-site pref.[a]	Photo pref.[a]	Comp.	Myst.	Spac.	Legib.
Vista	3.64	3.48	m	h	m	m
Obstructed	3.50	2.91	m	l	l	l
Lakelike	3.35	2.76	l	l	h	h

Note: l = low; m = medium; h = high.
[a]Somewhat different groupings.

Noteworthy Points

On-site preference study to avoid potential problems of photographs
Use of a sorting procedure to corroborate empirically based groupings
Similarity of two approaches

A VISUAL ASSESSMENT OF CHILDREN'S AND ENVIRONMENTAL EDUCATORS' URBAN RESIDENTIAL PREFERENCE PATTERNS

Augusto Q. Medina (1983)
Doctoral dissertation, University of Michigan (William Stapp, chairperson)

Overview. Medina used an information-processing approach to compare children's and environmental educators' familiarity with and preference for the urban environment. Though the study found that children share a common sense of place, comparison of these results with data from environmental educators showed that the two groups do not share the same perceptions of the urban setting. In addition to suggesting options for how environmental educators might approach teaching, the study offers both a methodology and a rationale for planners gathering input from children.

Pilot Study. To validate the assumptions underlying the study, Medina conducted a pilot study to see if children do share a common perception of urban settings; to check the use of photographs as a survey instrument; and to gain some sense of what in the city is salient to children. Thirteen students (grades 5–9; 10 boys, 3 girls; 11 whites, 2 blacks) sorted 58 photographs (of residential, industrial, and commercial scenes of Detroit and Ann Arbor) into groups. There were no criteria given for the sorting. Eight visual clusters emerged, even though the verbal descriptors for the groups used by the children were not related. This suggested that the children do categorize similarly and that there is value in using a visual instrument. The categories were: Well-kept Single-Family Homes; Rundown Single-Family Homes; Well-kept Multiple-Family Housing; Central City (cars, buildings, no vegetation); Car

Parking; Industrial Urban Wasteland; Urban Fringe/Natural Space; Parks/Recreation.

Photographs/Site. The final study used 56 black-and-white photographs, taken in the fall, of Detroit, Ann Arbor, and New York. Using the criteria generated from the pilot study, the scenes emphasized housing, commercial use, open space and recreation areas, and transportation. Streets, people, and water were avoided.

Sample/Survey. Respondents were 325 students (ages 12–14) from 11 classes in four Detroit schools. This sample was predominantly black. The survey consisted of a seven-page photo-questionnaire with two 5-point rating scales per photograph and background information related to students' neighborhoods, length of residence, type of housing, transportation options, and experience in other-sized communities. There were two treatment groups: 207 students rated each photograph in terms of familiarity with and preference for each photograph; 118, on how much elements seemed to "go together" (making sense) and how interesting the scene was (involvement). The making sense/involvement treatment was counterbalanced.

Ninety-two environmental educators responded to a mail survey asking for familiarity/preference ratings of the same photographs and for similar background information (plus their age).

Results

Background Variables. Both groups lived in similar neighborhoods (single-family housing), but the children came from more urban, developed areas with houses closer together and with less vegetation.

Familiarity. Ratings by the two groups were very different. Environmental educators were more familiar with 41 of 46 scenes. A comparison of the six most and least familiar scenes of each group suggested the two groups were familiar with different urban areas. Students had more familiarity with developed urban areas (concrete predominated in five of the six most familiar); the educators were most familiar with more "natural" scenes. Especially noteworthy: two of the students' least familiar scenes were open fields, one of which had been among the educators' most familiar. The three scenes of rundown, littered urban areas among the educators' least familiar did not appear in the students' least familiar set.

Preference. Although the two groups agreed on what they did *not* like, the educators' appreciation for nature (tree-lined streets, parks, fields) was not shared by the students, who preferred urban scenes (row housing, single-family residences, and commercial areas).

Categories. Nonmetric factor analysis (SSA-III) using the students' ratings generated eight patterns. Mean ratings for these are shown in Table B.11.

Familiarity. Again, the students had consistently lower familiarity with these patterns, even though they had the "home advantage" (60% of the photographs were from their hometown, Detroit). Medina noted that, "disturbingly," children were no less familiar with Rundown Urban than they were with Multiple-Family Housing or Tree-Lined Residential Streets, suggesting that children may have repeated contact with low-quality environments. The two groups also differed on the order of familiarity, differing most on Tree-Lined Residential Streets (more typical of high-income areas) and Retail City (taken in New York), two areas not common in Detroit. Without these, the ranking order for the two groups is similar, suggesting that students have less experience with

Table B.11. *Familiarity and preference means for both samples (Medina, 1983)*

	Familiarity		Preference	
	Stud.	Educ.	Stud.	Educ.
Urban Mobility	3.6	3.9	3.1	1.9
Single-Family Housing	3.3	3.6	3.3	2.7
Urban Parks	3.0	3.6	2.4	2.9
Multiple-Family Housing	2.8	3.2	2.8	2.3
Tree-Lined Residential Streets	2.7	4.2	2.7	3.8
Rundown Urban	2.7	3.1	1.3	1.6
Retail City	2.5	3.2	1.8	2.1
Industrial/Factory Sites	2.4	3.0	1.7	1.6

but are familiar with the same kinds of urban places as the educators.

Preference. There were major differences between the two groups in seven of eight patterns. Educators preferred scenes suggesting privacy and quiet (especially those with trees), whereas students had higher preference for two housing patterns (Multiple-Family and Single-Family) and Urban Mobility. The different preference order of the patterns showed the groups agreed only on what they did not like.

Familiarity as Predictor of Preference. Familiarity proved a good predictor of preference, especially for the educators (a positive linear correlation). Students had a two-tiered system that differentiated the home environment and the rest of the urban environment. Within these, the relationship between familiarity and preference was linear and positive. Students rated the home environment higher in preference relative to familiarity than other settings. Where opportunities for involvement are lacking, however (e.g., Rundown Urban), greater familiarity does not enhance preference.

Observations on the Patterns. The patterns generated suggest that (1) children's perception of the environment is rich and diverse. They seemed to organize and interpret the urban setting most importantly in terms of function (what there is that is interesting to do; e.g., high preference for Urban Mobility) and/or by form (the visual appearance). (2) Housing is not a unitary concept as indicated by the three types differentiated in the patterns; home is an important concept in the childhood experience. Medina interpreted preference and familiarity ratings for these patterns in terms of how they related to the children's urban experience; for example, students seemed to prefer more places with a variety of interesting things to do with friends.

Making Sense/Involvement. The high correlations of results between these variables and with preference (all > .90) from the second treatment suggested that making sense and involvement concepts as presented in this study tapped little more than the students' preferences. Medina concludes that, since the process of interpretation is rapid, unconscious, and automatic, sorting out the two may have been difficult. "With children, involvement and preference seem to be almost synonymous"; that which is seen as involving is seen as also making sense.

Noteworthy Points

Difference between perceptions of "experts" and others
Importance of familiarity as a predictor of preference
Suggests age difference in what environments are satisfying and preferred; children more interested in social environments and places high in opportunities for involvement; adults (educators) prefer more natural, quiet settings (although, here, age and ethnic group are confounded)

Although familiarity is a cue to preferred environments, in terms of children it is important to recognize that high familiarity may result from lack of better choices.

VISUAL PREFERENCE AND IMPLICATIONS FOR COASTAL MANAGEMENT: A PERCEPTUAL STUDY OF THE BRITISH COLUMBIA SHORELINE

Patrick A. Miller (1984)
Doctoral dissertation, University of Michigan (Rachel Kaplan, chairperson)

Overview. Miller's research focuses on ferry passengers' perception of various shoreline landscapes and compares these preferences with those predicted by professional planners. Results show that people do care about shoreline appearance and that a number of assumptions made by planners about these concerns are incorrect. Miller suggests ways of incorporating scenic values to make planning for shoreline development more comprehensive.

Site/Photographs. The 62 color slides used represented the diversity of landscapes found along the British Columbia shoreline between Victoria and Prince Rupert. Seven categories of land use were represented: industry, timber harvesting, log booming, shoreline roads, nonindustrial development, natural bald spots, and undeveloped shoreline.

Sample/Survey. The survey was conducted on board the British Columbia ferries. The sample of 476 volunteers represented a fairly broad cross-section, including both locals and tourists. During the ferry trip, the respondents listened to taped instructions, observed slides, and filled out a three-part written questionnaire about (1) their preference for each scene (5-point scale); (2) labeling or naming 10 preselected scenes from the original set; and (3) background information. (Three pretests were used to determine appropriate viewing time spans and the clarity and length of the audio instructions.)

Nine resource managers and planners ("experts") viewed 26 scenes drawn from the same set of slides and were asked (1) to anticipate the groupings of these scenes by the ferry passengers, (2) to suggest how different subgroups might respond to the pictures, and (3) to discuss their reactions to the ferry survey results and the current focus in planning on visual concerns.

Results

Preference. The categories generated with SSA-III reflected various degrees of human influence:

Natural with Depth (mean preference 4.15)
Small Structures in Natural Settings (3.92)
Low Flat Shorelines (3.19)
Clearcuts and Natural Bald Spots (3.19)
Log Booming (3.04)
Shoreline Roads (2.56)
Intensively Developed (2.06)

Distance and Preference. For five of the categories, there were suitable scenes (far and near vantages) to permit analysis of the influence of distance on preference. Preference for highly preferred (natural) scenes tended to decrease with distance; preference for less preferred (more human-influenced) ones increased with distance. Miller concluded that distance does have an effect on preference but that it is not large unless the distance is very great.

Demographic Characteristics and Preference. Of the demographic characteristics studied (age, income, place of residence, occupation, and reason for taking the ferry ride), only for age were there consistent significant differences. Older respondents had higher preference for all the scenes but especially those most influenced by humans.

Experts' Predictions of Preference. The planners had trouble separating their own reactions from their expectations when they attempted to predict the groupings. They could anticipate extremes in preference (i.e., what people would like most and least) but were not very successful at predicting middle preferences. They also anticipated many more differences on the basis of demographic characteristics than were manifested in the results. Miller concluded that, as experts, these people had a different perception, which unfortunately did not help them understand the public any better.

Noteworthy Points. The consistency of the sample's preferences serves to dispel some preconceptions about certain subgroups; for example, the assumption that those with higher incomes have higher preference for nature.

Nature with depth category again indicating the importance of Mystery in preference
Expert inability to speak for public - importance of public input

Although the intensity of development affects preference (more intense, lower preference), that intensity is a function of the relationship of the activity to the environment in terms of (1) its size, (2) how it covers the landscape (i.e., whether vegetation is dispersed through the area); (3) its transiency/permanency; (4) what the activity is; (5) how it is spatially arranged.

Higher preference for landscapes that have the appearance of or offer the potential for refuge, shelter
People more familiar with the landscape could differentiate clearcuts from naturally occurring bald spots; tourists did not

A VISUAL PREFERENCE STUDY OF URBAN OUTDOOR SPACES IN EGYPT

Laila Stino (1983)
Doctoral dissertation, University of Michigan (Charles Cares, chairperson)

Overview. Stino examined the experience and preference for different kinds of urban space in Egypt. Special consideration was given to the role of natural elements in urban space and the influence of group differences and culture. Results suggest that, in addition to function, preference for open spaces is influenced by the presence of nature. This preference, however, is not unidimensional; the spatial array of these elements is critical. Though water and mature trees were included in the most preferred photographs, scenes with excessive vegetation, with vegetation that obscured the view of the street or that blocked out sunlight, were not preferred. Recommendations included design sketches based on the study results.

Site/Photographs. The 42 photographs represented seven spatial environments as defined by function (e.g., active streets, squares, avenues and boulevards, walkups). The major difference between the city photographed, Heliopolis, and study site was Heliopolis's more mature vegetation.

Sample/Survey. In an at-home interview, 103 residents of Madinat Nasr, a community near Cairo, were asked to rate photographs in terms of how much they liked each (on a scale of 1 to 10). They also responded verbally to 16 questions about plant materials (appreciation of, care, park use), about the neighborhood (use patterns, sense of safety, and privacy), and about the people (family structure, neighborhood characteristics).

Results Relating to Nature/Preference. The most preferred scenes (preference 3.9–4.17 on a converted 5-point scale) were those (1) characterized by trees, shrubs, and/or water; (2) in which the space was well defined; and (3) that offered opportunities for diverse activities. The least preferred were "crowded" and not orderly.

Categories. SSA-III produced categories similar to those first used in selecting the photographs. Table B.12 shows these categories and the mean ratings. The emergence of two separate Walkup groups differing in terms of vegetation was unexpected. Apparently, the basis for dimension definition involved more than simply function - natural elements and spatial definition also play roles. The presence of shade trees and water was especially important, both as nature components and in defining spatial boundaries, providing refuge, and restoring human scale to the urban setting.

Table B.12. *Categories and mean preference ratings (Stino, 1983)*

A priori	Empirical categories	Pref.
Squares	Squares	4.05
Avenues & Boulevards	Public Areas	3.89
Walkups	Vegetated Walkups	3.74
	Nonvegetated Walkups	3.26
Shelters	Shelters	3.63
Active Streets	Active Streets	3.42
Small Residential Sts.	Small Residential Sts.	3.41
Shopping Sts. & Market	Shopping Sts. & Market	2.96

Noteworthy Points

Use of public input to guide design ("expert") recommendations
Evidence of cross-cultural differences (preference for Walkups over Small Residential Streets) highlights importance of knowing one's design public

SCENERY AND THE SHOPPING TRIP: THE ROADSIDE ENVIRONMENT AS A FACTOR IN ROUTE CHOICE

Roger S. Ulrich (1974)
Doctoral dissertation, University of Michigan (Gunnar Olson, chairperson)
[Also available as Michigan Geographical Publication No. 12]

Overview. Ulrich assessed the importance of non-economic variables affecting the route chosen for a "functional" (shopping) trip. He hypothesized and showed that a purely "economic view" of route behavior used by geographers (i.e., that people will always minimize travel time and distance) was unrealistic and that other factors, especially roadside scenery, were important in route choice. The study also presented empirical data for planners, both in terms of roadside preferences and supporting the importance of recognizing that people have a broad range of behaviors more appropriately met by offering a diversity of route options.

Setting Sample. The study focused on route choice of two parallel roadways offering a community relatively equal access to a shopping center in Ann Arbor, Michigan. The parkway was more scenic but slower; the other, the expressway, was faster and less scenic. A subdivision of about 400 relatively homogeneous, middle-to-upper-income families provided the study population for both a preliminary and a final study. For the preliminary study, 31 respondents (11 men, 20 women) were drawn by a stratified sample using the Ann Arbor telephone directory. Forty-eight usable surveys (24 men, 24 women), derived by the same procedure, constituted the final study sample.

Photographs. One part of the study used 53 black-and-white photographs (including some of the two roadways under study) representative of southeastern Michigan roadside scenery. A five-person panel rated 127 scenes using a "free category sort," whereby they sorted the pictures in terms of what visually "went together." Based on these results, all water scenes were omitted, and the judges analyzed the remaining scenes in terms of distance/space, complexity, and coherence. The final 53 photos used represented a balance of these criteria.

Survey

Preliminary Study. This open-ended interview explored route choice of 31 shoppers. Their responses verified that shoppers did have knowledge of both routes and that non-economic conditions were factors in their route choice. Several variables emerged that were explored further in the final study.

Final Study. The survey included a structured interview format, conducted in the respondent's home. The sample drawn had been contacted first by letter, then by a follow-up phone contact to arrange the interview time. Residents discussed their shopping patterns and responded to the relative importance and advantages/disadvantages of questions (6-point rating scale) dealing with the roadside environment, driving speeds, the roads themselves (curvy, hilly, straight, etc.), number of stoplights, as well as the shopper's mood and needs (e.g., in a hurry or not, safety, etc.). The second part of the interview used photographic preference ratings to identify potential nonverbal issues related to landscape preference. Participants rated 53 photographs of natural roadside views on a 6-point rating scale.

Results

Questionnaire. All respondents were aware that the expressway route was faster, but 56% of trips were on the parkway. Stratifications by sex, rate of parkway use, and whether or not one had children showed some differences, but shopping frequency and age of vehicle were not significant factors in route choice. In spite of marked difference on other issues, there was consistent agreement regarding the scenic experience of the parkway. Category identification methods, SSA-III and ICLUST, generated eight categories representing independent groups of attitudes underlying route choice behavior: Route Preference under Different Conditions; Expressway: Danger/Safety and Attention; Scenic Attributes of Parkway; Parkway Hills and Curves: Negative Aspects; Mood: Not in a Hurry, Want to Relax; Mood: In a Hurry; Parkway Driving Speed; Parkway Attentional Level. Interestingly, no category emerged that related to economic (distance-minimizing) variables, whereas one of the categories that emerged dealt specifically with the scenic quality of the parkway "view." The parkway scenery received the highest rating of all factors, whereas that of the expressway was rated the most important disadvantage.

Photographic Preferences. Categories generated from the roadside scene photographs, using SSA-III (loading > .45) and ICLUST, were (in order of declining preference):

Parklike, Parkwaylike (4.75) – mowed grass and trees; feeling of distance and space

Foreground Scenes (4.14) – foreground screens with sense of open space in middle ground

Uniform Texture, Tall Weeds (3.31) – unmowed vegetation in fore/middle ground, low complexity

Raggedness: High Complexity (3.30) – complexity in foreground, "scruffy," low in coherence

Low Complexity: Mowed Weeds (2.97) – mowed weeds in foreground, low in complexity, restricted sense of space

All parkway route pictures loaded on the most preferred Parklike, Parkwaylike category, indicating the parkway is seen as a scenic amenity. An examination of group differences revealed no significant differences.

Noteworthy Points. This is one of only a few studies to offer empirical evidence demonstrating benefits of a scenic roadside amenity. Traditional views of people as "economic man" would have placed emphasis on the minimization of travel time. Instead, the study suggests that noneconomic factors, including safety, attention demands, and mood, as well as the visual environment, are important.

COMPARISON OF MEDIA FOR PUBLIC PARTICIPATION IN NATURAL ENVIRONMENTAL PLANNING

William W. Weber (1980)
Doctoral dissertation, University of Michigan (Rachel Kaplan, chairperson)

Overview. Weber compared three forms of visual media (video, color slides, and graphic representation) for their effectiveness in conveying information about natural environments. Aerial imagery of three rivers was presented as it might be for a public presentation to assess the rivers' conservation value. Color slides proved most effective in eliciting positive impressions of the importance of preserving river segments. Cognitive involvement was enhanced with the use of graphic imagery.

Setting/Media. Matching aerial views of parts of three rivers, two in Illinois, one in Michigan, were depicted in video, color slides, and graphic representation (topo maps, sketches, satellite images). The slides and video were taken at the same time to match atmospheric conditions and perspective. For each river, the edited video was 2 1/2 min long; there were 14–16 slides of each river and 10–12 graphic representations.

Sample/Survey. The 113 individuals were recruited on the basis of expressed interest in rivers and included students, members of environmental groups, anglers, and so on. Participants were randomly assigned to one of three media conditions. The audio was the same in all cases. The survey had three sections: (1) ratings of various environmental attributes relating to the naturalness of the three selected rivers – (a) how participants thought experts would rate that characteristic, (b) how much the participants liked the characteristic, (c) what conservation priority rating seemed appropriate; (2) preference ratings (1–5 scale) of 20 slides of a number of different rivers and environments; (3) demographic information – age, interests, occupation, experience with various media.

Results

Media Effectiveness. Three categories were developed (based on $r > .40$): Experts' Naturalness, Personal Naturalness, and Conservation Rating. Comparisons of presentation methods for these categories showed preferences generally high for color slides and lowest for video. Weber postulated that video, like television, tends to make the scenes seem "commonplace" so that viewers were less impressed with the scenes' naturalness and, consequently, its conservation as a resource. He emphasized that both the medium and the message determine effective communication and that some subgroups of the audience (here, conservation group members) may be more sensitive to some of the media techniques.

Environmental Preference. Dimensional analysis of preference ratings for the 20 river scenes (using ICLUST and SSA-III) yielded three categories: Natural, Undisturbed (mean preference 4.03), Regrown, Moderately Disturbed (3.30), and Agriculture, Disturbed (2.42). Preferred elements included dense vegetation, absence of human habitation, water, and trees. Nonpreferred aspects were row-crop agriculture, prominent concrete bridges, erosion, and exposed riverbanks. The prominence or concealment of human development by natural features influenced preference, suggesting that human elements can exist in preferred settings. One surprise was the low preference for Agriculture as a distinct element. Mystery, especially as characterized in the Natural, Undisturbed category, enhanced preference.

Demographs. The differences in perception of individuals of different ages, students, and recreationists reinforced the notion that a single aesthetic or visual standard is not appropriate. Although there was no discernible pattern, background did affect evaluation scores. Those with less recreational or environmental policy experience gave higher ratings and conservation priority to the more developed and everyday natural aspects of the river segments than did those more politically involved. The less active also tended to put more faith in the experts' judgments.

Noteworthy Points

Use of environmental preference as a predictor of evaluative ratings and as a valuable, nonpolitical input for public policy decision making
Analysis of different media as tools for conveying information about natural environments
Support for the value of everyday nature, especially by the less environmentally active public

A FUNCTIONALIST APPROACH TO ENVIRONMENTAL PREFERENCE

David Woodcock (1982)
Doctoral dissertation, University of Michigan (Stephen Kaplan, chairperson)

Overview. Woodcock compared preferences for three distinct biomes (rain forest, savanna, mixed hardwoods) and examined the effectiveness of a series of predictors based on Appleton's Prospect/Refuge theory and on the information-processing approach. Using a larger sample of scenes than had been the case in previous biome preference research, he found mixed hardwoods to be the most preferred biome. Among the predictors studied, Prospect and Mystery were particularly effective; Refuge, surprisingly, was negatively correlated to preference.

Photographs. Seventy-two slides representing three biomes (savanna, rain forest, and hardwood forest) were rated by a panel of three judges. The judges evaluated the scenes in terms of six affordances: Primary Prospect, Secondary Prospect, Primary Refuge, Secondary Refuge, Mystery, and Legibility, using a 5-point scale.

Definition of Predictor Variables

Legibility - slide readily interpretable and coherent; offers a few strong features or a few distinctive regions whose configurations are easily memorable; scenes low in Legibility offer a sample of shapes, textures, and forms
Mystery - slide suggests significant changes in landscape just beyond the immediate view
Primary Prospect - slide affords view of surrounding landscape for miles around; generally, view taken from high vantage point
Primary Refuge - slide shows evidence that it was taken from a landscape feature affording concealment
Secondary Prospect - slide includes good vantage point
Secondary Refuge - slide includes landscape features that promise good concealment

Sample/Survey. Two hundred college students (102 female, 98 male) rated each slide (using a 5-point scale) in terms of preference and how difficult it was for them to come to that decision. The slides were presented in sets of 24. The sets were presented either by biome or as a random mixture using eight slides from each of the three biomes. Slide order was counterbalanced to reduce order effects.

Respondents filled out an Environmental Preference Questionnaire and Environmental Experience Questionnaire. The first had respondents rate their reactions to certain verbal descriptions of environmental settings on a 5-point scale. The second was designed to assess lifetime experience with different biome types and urban settings.

Results. Major groupings created by ICLUST and nonmetric factor analysis (MVA) were quite comparable: Savanna (mean preference 3.06), Dense Hardwood Forest with underbrush (3.04), Open Hardwood Forest with open ground (3.73), and Rain Forest (2.83).

Predictors of Preference: Affordances. The predictive power of each of the six affordances was examined using simple correlations and multiple linear regression. Each analysis was carried out separately for the three biomes (see Table B.13). The unexpected relationship of Primary Refuge and preference led Woodcock to propose several other possible predictors, Agoraphobia and Claustrophobia (as types of setting aversion - to open areas and closed spaces); Edge Quality and Edge Attraction (as types of preference - for light when in darkness and for cover when in the open). He also examined the influence of Aridity and Ease of Locomotion. Only Agoraphobia emerged as a potentially useful measure.

Aesthetic Ratings. A judge, well schooled in theoretical aesthetic theories, rated the slides on how well they fulfilled three criteria: Sublime, Beautiful, and Picturesque. The relationship between these predictors and preference was significant but weak compared to seven of the other predictors examined.

Individual Differences: Sex. Woodcock related differences in preference by sex to evolutionary role differentiation (males as hunters, women as gatherers). MVA and ICLUST revealed similar differences in preference patterns: Males differentiated the scenes into two protypes: the more preferred savanna and the less preferred rain forest. (Woodcock suggests one would expect the hunter to prefer open country and vistas.) Females did not see this dichotomy but tended to center their organization around forest views (where he suggests it would be "safer" to do most of their food gathering). Females had lower preference for scenes characterized by lack of cover (e.g., savannalike). Mystery played a significant role only in male preference.

Previous Experience. The significantly higher familiarity of the sample with hardwoods suggests a strong relationship of familiarity with preference, although not enough

Table B.13. *Correlations of predictors and preference (Woodcock, 1982)*

Affordance	Savanna	Hardwood	Rain Forest
Primary Prospect	.44[a]	.55	ns
Secondary Prospect	ns[a]	.40	.36
Primary Refuge	ns	−.59[b]	ns
Secondary Refuge	ns[a]	ns	ns
Mystery	ns	.55[a]	.56[a]
Legibility	ns	.48	.40

[a]Predicted by multiple linear regression.
[b]Woodcock considers this to be not significant because it is opposite the predicted outcome.

respondents had had experience with other environments to make other comparisons possible.

Noteworthy Points

Comparison and synthesis of various preference theories (evolutionary and aesthetic)
Extensive review of habitat preference research in nonhuman species
Analysis of paleontological evidence concerning the type of environments in which human species evolved
Sex differences relating to preference for different landscapes, role of familiarity in preference

Ed. note. In subsequent analysis of the data, Legibility was found to be a significant but interactive predictor of preference.

A CROSS-CULTURAL COMPARISON OF PREFERENCE FOR KOREAN, JAPANESE, AND WESTERN LANDSCAPE STYLES

Byoung-E Yang (1988)
Doctoral dissertation, University of Michigan (Terry J. Brown, chairperson)

Overview. Yang's study is cross-cultural both with respect to the landscape styles that are studied and by virtue of the participants. In addition, three landscape elements (water, vegetation, and rock) are examined, as well as spatial layout. Both Korean and western participants preferred the Japanese landscapes, as well as the landscape styles distinct from their own culture. Similarly, water was highly preferred, and rock far less favored, by all groups. The perceptual categories, however, showed some differences between the Korean and western samples. Furthermore, the preference ratings were uniformly higher for the western participants.

Settings/Photographs. All but two of the possible combinations of the three landscape styles (Korean, Japanese, and western) and four landscape characteristics (water, vegetation, rock, and layout of space) were sampled, with four scenes representing each cell. (Western use of rocks and Japanese layout of space were excluded.) The 40 scenes were printed in a photo-questionnaire. All scenes were taken in Korea.

Ratings. A jury of landscape professionals reviewed the scenes and rated them in terms of style and characteristics, leading to the final selection of 40 scenes. The photo-questionnaire included four scenes per page with two ratings requested for each scene: preference and familiarity, each using a 5-point scale. A separate set of photo-questionnaires included five rating scales beneath each scene (degree of Complexity, Uniqueness, Legibility, Mystery, and Coherence) and also used 5-point rating scales. Two versions of each booklet were printed to control for order of ratings. Several background questions were also included in the booklets.

Sample/Survey. For the preference aspect of the study, the sample included three distinct groups. The Korean citizen sample consisted of a multistage random sample of 415 individuals. The Korean college students sample (N = 135) consisted of students at Seoul National University, including both a random sample and students in landscape professions. The western participants were all tourists in Seoul, approached at one of three historic palaces (N = 110).

Results. Table B.14 includes the mean preference ratings for each sample for the different landscape styles and characteristics. The Japanese-style scenes were significantly preferred to either other style by each sample. Each sample also significantly preferred the water scenes to the other landscape characteristics, and rock scenes were least liked. For the Korean sample, the western scenes were preferred to Korean scenes; for the other samples, these differences are not significant. The overall pattern of ratings is similar for both groups, although the magnitudes for the western sample are substantially higher in all instances.

Perceptual Categories. SSA-III and ICLUST procedures were used to determine preference-based categories for each sample. The resulting groupings had many similarities and some differences. The Korean-based categories were more homogeneous with respect to landscape characteristics, whereas the western-based categories followed

Table B.14. *Mean preference ratings of landscpe styles for each sample (Yang, 1988)*

Landscape element	Korean sample				Western sample			
	Korean	Japanese	Western	Total	Korean	Japanese	Western	Total
Water	3.16	3.47	2.91	3.18	3.47	4.06	3.18	3.57
Rock	2.23	2.62	—	2.42	2.87	3.21	—	3.04
Vegetation	2.66	3.16	3.09	2.97	3.13	3.75	3.16	3.35
Layout	2.48	—	2.67	2.58	3.43	—	3.09	3.26
Total	2.63	3.08	2.89	2.84	3.23	3.67	3.14	3.34

somewhat more closely along landscape style. Thus the Korean grouping Water Surrounded by Vegetation and the western-based Japanese-style Landscape had six items (of eight and seven, respectively) in common and in each case was the most preferred category. The least preferred Korean category, Rock with Sparse Vegetation, had nine scenes in common with the least preferred western-based Korean-style Landscapes. The other two perceptual categories for each culture reflected differences between formal and informal (mostly western) styles. The informal category was preferred to the formal for each sample.

Noteworthy Points. There has been little cross-cultural research in preference involving nonwestern groups. Furthermore, this study is the first to examine differences between oriental landscape styles. The strong similarities across cultures are striking. The strong preference for water scenes is not as surprising as the consistently low ratings for the rock landscapes. The dissertation also includes an excellent summary of the history and characteristics or different landscape styles and design recommendations based on the research.

APPENDIX C

OUTDOOR CHALLENGE PROGRAM

Appendix C consists of two parts. The first includes a sampling of the research instruments that were used in the program and are pertinent to the discussion in chapter 4. The second is an annotated bibliography of the reports and publications that ensued from the program.

RESEARCH INSTRUMENTS

Over the course of the 10 years of this research program we used a great variety of open-ended and structured questions as well as various other tasks. A sampling of these, especially those pertinent to the presentation in chapter 4, are included in this appendix.

Participants completed information immediately before departure on the trip and at the conclusion of the trip; in addition, questions about solo were included both before and after that experience. The specific questions varied to some degree over the years, and the inclusion of other questionnaires also differed from year to year. The appended material also gives a sense of the context we provided for the participants in responding to these materials.

Some Items from 1973 Project

Please evaluate your skills with regard to the following activities, where 1 = practically none; 2 = very little; 3 = not too much; 4 = a moderate amount; 5 = a fair amount; 6 = a great deal.

knowledge about the woods
outdoor cooking

map reading
fire building
using a compass
finding your own food in the woods
setting up camp
making new friends
ecology
getting along with strangers in confined situations
rock climbing
long hikes
first aid
canoeing
How would your best friend describe you (aside from physical characteristics)?
What sort of things have given you the greatest sense of accomplishment or pride?
If you could change yourself in any way, in what way would that be?

The Process of "Getting to Know" (based on 1975 and 1976)

We would like to have your help in figuring out what kinds of things help one feel at home in an area like this. If we knew more about this, we could be more helpful to people who are not at all comfortable in the woods. For instance, if people could be given some general knowledge about woods and wilderness before going there, their experience might be more positive right from the start. But what sort of general knowledge would be useful?

We do know a few things about this process of "getting to know." One thing we know is that once you are in such a setting, the process goes really fast. A lot of things that are confusing and worrisome at first are no longer problems by the next day. But then, some different things can become a problem that were not in the beginning. We also know that it is really hard to figure out what it is that one feels one knows better.

To get your help with this we have prepared some questions that we would like you to answer. Since we know the process is so fast, we want your answers each day for the first few days and then again at the end of the first week. In addition to our questions, we would greatly appreciate any other ideas or insights you have. It is fine to discuss these things during the day, while you are walking – try to figure out what goes on as you get to feel more at home. Are there landmarks you use, are there regions that help you find your way, does water make it easier or harder, how much of the time are you aware of the sun, etc.? Do please write down any of these things on the sheets. Many thanks.

1. How well do you think you know each of these things about this setting?
   (1 = not at all; 5 = very much)
   what plants and wildlife are in the area
   layout of the land
   what the other people in the program are like
   compass directions
   where to find drinking water, sources of food
   what we're going to be doing
   what this place looks like on a map
   other
2. How much do the following features help give you a sense of this place?
   Are these features ever misleading or a source of confusion?
   streams islands swamps trails roads lakes hills
   stars rock formations ravines or valleys
3. How much do these things worry you, or make you uncomfortable?
   weather getting lost solo having enough food
   being alone getting sick that someone might get hurt
   bugs sore and tired being away from home dangerous animals getting along with others
4. To what degree do each of these things help make you feel comfortable here?
   I can do what I want to do
   peace and quiet in the woods
   cooperating with others
   using a compass
   I feel I can trust my own judgment
   nothing to be afraid of in the wilderness
   noticing things, hearing sounds in nature
5. Please add anything else that you think helps to figure out what goes on in the process of getting to feel comfortable.

This question, completed both before departure and at trip's end, is the basis for Table 4.6 (1981 version)

How well does each of these items fit the way you are feeling today? (1 = not at all; 2 = a little; 3 = some; 4 = quite a bit; 5 = very much)

easily irritated at small things
content with life as a whole
realistic about myself
harried, hurried, and rushed
able and willing to change things in my life that I am not satisfied with
enjoy getting along on less
it feels great to be alive
physically active and full of energy
curious about things in general
anxious about daily activities, how things are going
self-confident
generally in control of things
relaxed
able to concentrate
fearful of danger
comfortable with myself
eager for ways to simplify my life
happy with a slow pace
responsible in accepting work, chores, etc.
readily irritated by other people
annoyed by things like noise, pollution, litter, crowds, etc.
in tune with nature
caught up in the "rat race"
enjoy dealing with problems
all's right in the world

Reentry

Upon leaving the program, participants were given a booklet. The 1980 version, for example, had the following cover page:

Just as we were interested in your reactions as you became accustomed to your wilderness surroundings, we would now like to know how you are feeling as you readjust to being back.

For the Next 4 Days. We would like you to write down *each* day the kinds of things that bother or annoy you, the things you notice around you, what you enjoy doing. Do things here seem different from before? Do you feel different from before? If so, how?

Also, as you spend time away from the Outdoor Challenge Program, what do you remember most about your wilderness surroundings? What seems most special, or important, or most vivid in your memory of that time? What things about living in a wilderness area seem most different from the way things are now?

Please feel free to write down as much as you can think of – add extra pages, if you like.

In a Week or So. We would like you to consider some topics that you'll find on the green sheets of this journal. AND THEN MAIL THIS BACK TO US in the envelope provided. Many thanks.

The booklet's first page asked for the date and continued:

What kinds of things do you notice? How do you feel about people you see and talk to? What do you notice about yourself – how you're feeling and reacting to things? How about your ability to deal with daily problems and jobs? How do you feel different from before? What do you remember most about Outdoor Challenge? What seems most special about your surroundings there? What was most different from the way things are now?

The green pages started off with the following:

As you write in this booklet for the last time, please see if the following topics help you think of anything else you can tell us, either about how you feel now, or about how things felt different during your wilderness trip.
noises, sounds around you privacy, quiet the sense of time passing eating, sleeping other people you are with planning ahead your energy, excitement exercise, using your body places (buildings, cars, the outdoors, etc.) tools, the things you use each day involvement, concentration, being absorbed in what you're doing personal responsibilities, chores, jobs

REPORTS AND PUBLICATIONS

Chapter 4 draws heavily on various reports and publications prepared since the early 1970s about aspects of the program. Included here is a listing of the major papers used in developing the chapter as well as a brief annotation of their content. (Reports were all submitted to USDA Forest Service, North Central Forest Experiment Station.)

R. A. Hanson (1973). Outdoor Challenge and mental health. *Naturalist, 24,* 26–30.
A description of the program in its initial form published in the journal of the Natural History Society of Minnesota. Other contributions to this issue, devoted to the McCormick Experimental Forest, described the history of this large property

given to the U.S. Forest Service for research purposes as well as the flora, fauna, and geology of the area.

R. Kaplan (1974). Some psychological benefits of an Outdoor Challenge Program. *Environment and Behavior, 6,* 101–116.
Results of the initial year of the research program. Though the article was based on a very small sample, the journal felt it made a useful contribution and might encourage more research in this area.

R. Kaplan (1974). *Project Outlook: Role of summer outdoor programs in interests and self-views.* Progress Report (16 pp.)
Summary of the second year of the research program, which entailed not only participants in Outdoor Challenge but high school students participating in various other programs as well as two control groups.

J. Frey (1975). *Getting to know the natural environment.* Progress Report (10 pp.). Co-operative Agreement 13-387.
Results of generic knowledge project for summer 1975, the first time structured materials were used in the "getting to know" research.

R. Kaplan (1977). Patterns of environmental preference. *Environment and Behavior. 9,* 195–216.
This publication contains comparisons of the various groups involved in the *Project Outlook* study (see above) at the initial data collection point. As such, it is a study of high-school-aged students and their interests and inclinations, independently of any particular summer programs.

R. A. Hanson (1977). An outdoor challenge program as a means of enhancing mental health. In *Children, nature, and the urban environment* (pp. 171–173). USDA Forest Service General Technical Report NE-30.
Part of a symposium presented in Washington, DC, in 1975. Hanson provides a description of the program and its intentions.

R. Kaplan (1977). Summer outdoor programs: Their participants and their effects. In *Children, nature, and the urban environment* (pp. 175–179). USDA Forest Service General Technical Report NE-30.
As part of same symposium, this provides an overview of the results presented at greater length in the *Project Outlook* report (see above).

S. Kaplan (1977). Tranquility and challenge in the natural environment. In *Children, nature, and the urban environment* (pp. 181–185). USDA Forest Service General Technical Report NE-30.
This paper explores the sense of clarity that people have when they experience a fascinating environment. (An expanded version of the paper, not specific to Outdoor Challenge, can be found in S. Kaplan & Kaplan, 1978, as "Attention and fascination: The search for cognitive clarity.")

R. Kaplan, S. Kaplan, and J. Frey (1979). *Final report: Assessing*

the benefits of a natural area experience; and *Orientation to a wilderness area experience.* Co-operative Agreements 13-451 and 13-452. USDA Forest Service, North Central Forest Experiment Station (73 pp.).
The document summarizes the program's research for 1974–78. Chapter 4 draws heavily on the description of the "getting to know" project and results of the solo experience. In addition, some of the insights about the impact of the program are also included in the chapter.

R. A. Hanson (1979). *Final report: A study of psychological factors influencing human carrying capacity within a natural environment.* Co-operative Agreement 13-423. USDA Forest Service, North Central Forest Experiment Station (23 pp.).
Hanson's report covers the same period (1974–78), with a focus on the impact of hearing and seeing others while on solo.

R. A. Hanson (1980). *Study plan: Maximizing wilderness benefits: Some methods for enhancing the process of a natural environment experience.* Co-operative Agreement 13-702. USDA Forest Service, North Central Forest Experiment Station (19 pp.).
A proposal for continued research on carrying capacity, providing a useful summary of the past efforts.

R. Kaplan (1982). *Sights and sounds during solo.* Progress Report (6 pp.).
Results of the social carrying capacity component of the project for the final 2 years of the program. These are summarized in chapter 4.

S. Kaplan and J. F. Talbot (1983). Psychological benefits of a wilderness experience. In I. Altman and J. F. Wohlwill (Eds.), *Behavior and the natural environment* (pp. 163–203). New York, Plenum.
The chapter not only provides presentation of many aspects of the research program but concludes with a theoretical analysis of components of a restorative experience. As such, this paper is central to both chapter 4 and chapter 6. (Table 4.3 and portions of Table 4.4 appeared in this publication.)

S. Kaplan (1983). A model of person–environment compatibility. *Environment and Behavior, 15,* 311–332.
A theoretical analysis that draws on the Outdoor Challenge findings to propose a more general model of what constitutes compatibility in a supportive and restorative environment. The paper plays a direct role in chapter 6.

R. Kaplan (1984). Wilderness perception and psychological benefits: An analysis of a continuing program. *Leisure Sciences, 6,* 271–290.
Based on the last 2 years of the program, this paper presents results using photographs (see chapter 3) as well as material on solo and the more general "methods and feelings." (Table 4.6 first appeared in this publication.)

J. F. Talbot and S. Kaplan (1986). Perspectives on wilderness: Re-examining the value of extended wilderness experiences. *Journal of Environment Psychology*, 6, 177–188.
A comparison of various aspects of the Outdoor Challenge Program to see the ways in which the shortened outing in the last 2 years might have affected the participants.

OUTDOOR CHALLENGE PROGRAM

APPENDIX D

BENEFITS AND SATISFACTIONS STUDIES

ABSTRACTS OF STUDIES FOR CHAPTER 5

The discussion in chapter 5, "Nearby Nature," draws heavily on the studies that are summarized in this appendix. The studies are marked with * when mentioned in the text.

The following studies are included in Appendix D:

Bardwell, 1985
Frey, 1981
R. Kaplan, 1985a
S. Kaplan, Talbot, & Kaplan, 1988
Talbot, Bardwell, & Kaplan, 1987
Talbot & Kaplan, 1984
Talbot & R. Kaplan, 1986

NATURE AROUND THE CORNER: PREFERENCE AND USE OF NEARBY NATURAL AREAS IN THE URBAN SETTING

Lisa V. Bardwell (1985)
Master's thesis, University of Michigan (Rachel Kaplan, chairperson)

Overview. Residents of a townhouse complex, with a wide variety of outdoor areas both within and adjacent to it, were interviewed about their perceptions and uses of nearby nature. Results indicated that the residents' yards were not generally preferred and were considered too small. On the other hand, public open spaces and a nearby woods were felt to be too big and uninviting, although they received much higher preference ratings than the yards.

Survey/Sample. Photographs depicting 14 nearby outdoor areas (such as yards, specific commons areas, areas within a nearby park, an adjacent woodlot, and a pond) were used in each interview. Residents were asked if they recognized each type of area and then rated each for frequency of use, preference, perceived size, and the adequacy of the area's size. Open-ended questions accompanied each of the ratings, so that the residents could clarify their responses. Respondents also answered questions about the amount of nature they had nearby, how important nearby nature was to them, and how well they liked living in the complex. Demographic information was also obtained.

Thirty-two residents were interviewed in their homes (representing roughly 10% of the dwelling units in the complex). The respondents were relatively young, transient, and most often childless.

Results

Preferences and Uses. The front and back yards received strikingly low preference ratings (2.7 and 2.8 on a 5-point scale). Preference ratings for the other areas varied from moderate to relatively high levels, with the highest ratings (3.9) given to two common spaces, the park entrance and an adjacent undeveloped field. Use frequencies and the types of uses reported were widely varied. The yards were used most frequently, although primarily in an indirect fashion, such as passing through them to the parking lot. The common spaces and the park were used relatively often, the natural areas least frequently.

Perceived Sizes, Adequacy of Sizes. The yards were perceived as extremely small and as being much smaller than people would like. The other outdoor areas were perceived as moderately large or quite large. The common areas in the complex and the adjacent undeveloped areas felt slightly too small to people, and all the areas within the nearby park felt slightly too big.

Interrelationships among Variables. When all 14 areas were combined, a high correlation was found between perceived sizes and preferences. However, if the yards were excluded, there was no clear relationship between perceived sizes and preferences for the remaining areas. Similarly, size adequacy was related to preferences for the yards but not for the other areas studied. Use frequencies were not related to preferences for the different areas. The relationship between use frequencies and size perceptions was also examined. Frequent users perceived the undeveloped open field and a scrubby area as being smaller than did infrequent users. For three areas, there was a significant relationship between use frequencies and size adequacy: Front yards (which everyone considered too small) were rated as less inadequate by frequent users than by others; the undeveloped field and the pond area in the park were considered too small by frequent users but were considered to be either of adequate size or too big by residents who used them less frequently.

Noteworthy Points. This complex was touted as an innovative "human-oriented" approach to townhouse design and won a Sensible Growth Design and Planning Merit Award in 1976. However, the decision to provide virtually no front yards and very small back yards (10–12′ × 12′) clearly was not seen as human-oriented from the residents' viewpoint.

Two general categories of uses of nearby nature were distinguished in the analysis of these patterns: Circumstantial (routine, necessary) and Intentional (voluntary, choosing a setting for favored activities or involvements).

Regarding the nonprivate spaces, the same area can feel too small for some individuals but feel too big to others. In some cases, frequency of use affected these perceptions.

Ed. note. Portions of this study are discussed in Talbot, J. F., & Bardwell, L. V. (1989), *Making "open spaces" that work: Research and guidelines for natural areas in medium density housing.* In G. Hardie (Ed.), *Changing paradigms.*

PREFERENCES, SATISFACTIONS, AND THE PHYSICAL ENVIRONMENTS OF URBAN NEIGHBORHOODS

Janet E. Frey (1981)
Doctoral dissertation, University of Michigan (Rachel Kaplan, chairperson)

Overview. Residents of 38 neighborhoods completed a survey dealing with nature interests and activities as well as neighborhood and life satisfaction levels. The results showed that nature accesss and nature-related involvements affect neighborhood satisfaction and also that neighborhood satisfaction levels affect feelings of general life satisfaction.

Survey/Sample. The survey included questions about individuals' nature-related activities and interests and about where these were pursued. It also asked how satisfied residents were with their neighborhood and with life in general. (The photo-questionnaire aspect of the study is not included in the discussion here.)

The survey was completed by 369 residents of 38 Ann Arbor neighborhoods (response rate, 42%). The neighborhoods were selected in terms of proximity to public parks, and both park access and other measures of nearby nature and open space were coded for each area.

Results. ICLUST and SSA-III analyses were used to establish categories for Nature Access, Nature Involvement, Neighborhood Satisfaction and Life Satisfaction ($p < .02$ was used as criterion of significance).

Nature Involvement. Two sets of items measure the participants' degree of Nature Involvement. The first, Special Nature, listed six types of nature areas and asked whether there were any near where the participants lived that were "especially important or enjoyable" for them.

The second set of items asked how frequently the participants pursued certain nature-related interests. This set combined into five groups: Nature Watching (watching birds and squirrels, noticing trees and others' gardens, taking a walk); Gardening; Picnicking; Wildlife Activities (feeding birds and squirrels); and Sports. Responses both to the Special Nature items and to the other outdoor involvements reflected high levels of interest in nearby nature. Sports was the least frequently pursued outdoor involvement.

Nature Access. This topic comprised five distinct categories, reflecting how close to home certain nature-related opportunities were located: Casual Nature (places to watch birds and squirrels, or to sit outside and relax); Picnic Spots; Nature Focus (places to collect leaves, take nature photographs, or notice trees or others' gardens); Special Nature (settings where a favorite tree or "special" outdoor spot was located); and Outdoor Recreation. Casual Nature settings were located closest to home for the participants, and Outdoor Recreation places were the farthest away.

Neighborhood Satisfaction. The satisfaction items formed five categories: Services; Outdoor Recreation; Trees; Outdoor Solitude; General Activity (variety and friendliness of people, variety of buildings, activity level, and availability of gardening places); and Lack of Problems (level of traffic, noise, litter, recent changes, feelings of security and safety).

Factors Affecting Neighborhood Satisfaction. Neighborhood satisfaction was strongly affected both by individuals' Nature Involvement and by their Nature Access. People living in heavily treed areas also were more satisfied with their neighborhoods, as were people for whom certain nature elements in their neighborhood seemed unique and especially valuable (based on answers to open-ended question).

Of the six Nature Involvement groups, only Sports did not enhance Neighborhood Satisfaction. And of the six groups of Neighborhood Satisfaction items, only the Lack of Problems cluster was unaffected by an individual's nearby-nature involvements.

All six Neighborhood Satisfaction clusters were affected by the Nature Access measures. Interestingly, Outdoor Recreation, like Sports, showed the least relationship to satisfaction levels. In contrast, those who lived closest to Special Nature settings reflected the highest ratings on all six groups of Neighborhood Satisfaction items.

There was a direct relationship between tree density and Neighborhood Satisfaction levels. People living in areas with the most trees reflected the highest satisfaction levels, whereas people in areas with fewer trees reflected lower satisfaction. The lowest Neighborhood Satisfaction was found among people living in "mixed" areas (neighborhoods that included contrasting portions, some with very large trees and others with very few trees). Higher Neighborhood Satisfaction was also found among participants who mentioned nature in answering an open-ended question concerning what they would miss if they moved away from their current neighborhood. Participants mentioned particular trees, wooded areas, wildlife, parks, views, and the landscape in general when answering this question.

Life Satisfaction, and Factors Affecting It. CIM procedures resulted in four Life Satisfaction categories: Social Satisfaction; General Satisfaction; Current Satisfaction; and Perceived Stress. Each of the six Neighborhood Satisfaction categories significantly affected individuals' Life Satisfaction levels. Only Perceived Stress was unaffected by any of the Neighborhood Satisfaction categories.

Noteworthy Points. Both the availability of nearby nature and the individual's degree of involvement with nearby nature contribute strongly to Neighborhood Satisfaction levels. Sports activities and Outdoor Recreation opportunities, however, had minimal impacts. Again, nearby nature has strong effects, but the critical elements are small-scale settings and elements rather than large open spaces.

Rather than emphasizing park access and opportunities to participate in sports, public officials should demonstrate an increased concern with providing and maintaining trees and with creating a variety of natural settings both in and near urban neighborhoods. These are the elements that have been shown to affect individuals' well-being.

Neighborhood Satisfaction affected Life Satisfaction levels.

Ed. note. Portions of this study are discussed in:

Talbot, J. F. (1982). Zoning reconsidered: The impact of environmental aesthetics in urban neighborhoods. In P. Bart, A. Chen, & G. Francescato (Eds.), *EDRA 13: Knowledge and design* (pp. 154–159). Washington, DC: Environmental Design Research Association.

Talbot, J. F. (1988). Planning concerns related to urban nature environments: The role of size and other physical features. In J. L. Nasar (Ed.), *Environmental aesthetics: Theory, research, and applications.* New York: Cambridge University Press.

NATURE AT THE DOORSTEP: RESIDENTIAL SATISFACTION AND THE NEARBY ENVIRONMENT

Rachel Kaplan (1985)
Journal of Architectural and Planning Research, 2, 115–127

Overview. Residents of multiple-family complexes answered a questionnaire dealing with their access to nearby nature, their preferences and uses of such areas, and their satisfaction with their housing situation. Responses revealed that views of nature elements, particularly of trees and woods, were extremely effective in enhancing residential satisfactions. The view of a garden played a special role in enhancing people's satisfaction with Community, which was unaffected by other variables reflecting the physical environment. The perceived adequacy of nearby nature also affected satisfaction levels, whereas variables reflecting the dominance of constructed elements in the view reduced satisfactions. Having views of and adequate access to large open spaces and parks had no perceived impacts on residential satisfactions.

Survey/Sample. The survey included questions on how dominant certain elements (both natural and built) were in the view from the dwelling; how far away different kinds of nature areas were located, how adequate these were for individuals' needs and how often they used them; how many trees were located close to their dwelling; how satisfied they were with their apartment complex; and how satisfied they were with their lives in general. The questionnaire also included photographs of outdoor scenes which were rated for preference and for availability near the respondent's dwelling.

The survey was distributed to residents of nine apartment complexes in Ann Arbor. The response rate was 32% (sample size, 268). The participants were relatively young, transient, and most often childless.

Results

Neighborhood Satisfaction. ICLUST and SSA were used to identify four categories of Neighborhood Satisfaction. Physical Facility (security and safety, maintenance, parking); Size (number of people, number of children); Community (variety and friendliness of people, sense of community); and Nature (amount of open space, private open space, trees and landscaping, recreation opportunities, and having enough nature nearby).

Predictors of Neighborhood Satisfaction. Having a dominant view of a well-landscaped area enhanced Physical

Facility and Nature-related satisfactions. Views of small trees, of large trees, and of woods enhanced these two satisfaction categories and Size satisfactions as well. The same three categories were also enhanced by feelings that the individual's needs for a natural area and for a place for taking walks were adequately met, and by having a greater number of trees close to one's dwelling. Satisfaction with Community was enhanced by having a garden as a dominant element of the view from the window (regardless of whether or not individuals worked in gardens themselves), and also by perceptions that individual needs for a natural area, for a place to take walks, and for a place to grow things were adequately met. Having a dominant view of a park did not significantly enhance satisfaction levels. Feeling that one's needs for access to a park were adequately met also did not affect satisfaction levels.

Having dominant views of certain built elements (such as parking, power lines, busy streets and highways) resulted in significantly lower satisfaction levels on all groupings except Community satisfaction.

Noteworthy Points. Urban nature settings separated into three distinct categories, in terms of their effects on residential satisfactions: (1) gardens (which affect Community satisfaction); (2) nature elements and natural settings (which affect all other, more broadly based satisfactions); and (3) large parks and open grassy areas (which had no discernible impacts on neighborhood satisfactions).

Community satisfaction has a different dynamic than the other satisfaction categories. Gardening affects this, which in this context is more social, whereas other physical elements (both natural and built) do not.

Adequacy was a better predictor of satisfactions than proximity (the conceptual "othereness" aspect makes more of a difference than the purely physical conditions). Adequacy ratings were the only variables that affected all categories of residential satisfaction including Community.

The findings show that many of the desirable qualities of single-family living are attainable in the multiple-family context through the judicious use of nature.

Ed. note. More detailed presentation of this material is in R. Kaplan (1981), *Nearby nature and satisfaction with multiple-family neighborhoods,* Report to the Planning Department, City of Ann Arbor, MI (35 pp.).

Portions of this study are also included in:

Kaplan, R. (1982). Managing greenspace in multiple-family neighborhoods. In *Proceedings of the 1982 Convention of the Society of American Foresters*. SAF Publication 83-04, Library of Congress No. 83-60910.

Kaplan, R. (1983). The role of nature in the urban context. In I. Altman and J. F. Wohlwill (Eds.), *Behavior and the natural environment*. New York, Plenum.

Discussion here focuses on benefits and satisfactions. Appendix B focuses on photo-questionnaire aspect of the study.

COPING WITH DAILY HASSLES: THE IMPACT OF NEARBY NATURE ON THE WORK ENVIRONMENT

Stephen Kaplan, Janet F. Talbot, and Rachel Kaplan (1988)
Project Report, U.S. Forest Service, North Central Forest Experiment Station, Urban Forestry Unit Cooperative Agreement 23-85-08

Overview. Three groups of employees completed surveys concerning their perceptions of both job stresses and the value of brief "restorative" involvements that might help relieve stresses during the workday. They also answered questions about their general health and their feelings of overall well-being. The results indicated that job stresses exist and that they significantly affect employees' physical and mental well-being. At the same time, various workday involvements, many of them nature-related, were considered at least moderately restorative by the participants. In addition, both the individuals who indicated they could see outside from their desks and the others who worked outside reflected less job stress and exhibited fewer problems than the rest of the sample.

Survey/Sample. The survey included questions about perceived job stresses and about the value of various settings and activities as workday restoratives. It also included items on potentially restorative experiences outside of work as well as items reflecting levels of general well-being, physical health, and overall life satisfaction. Some demographic information was obtained, and respondents also answered questions concerning their degree of contact with the outdoors during the workday.

The sample consisted of 168 employees of two public agencies and one large corporation, all located in Washtenaw County, Michigan.

Results. ICLUST and SSA-III analyses were performed to establish clusters of Job Stresses, Restorative Involvements, Physical Health, and General Well-being. (Predictive relationships significant at $p < .05$ are reported here.)

Perceived Job Stresses. Employees reflected considerable degree of job stress. They felt that their workloads were fairly demanding, that they had to deal with a considerable degree of interference in their daily tasks and also with frequent changes in management priorities. Overall, respondents felt moderately pressured and frustrated by their jobs. The results also indicated that employees held certain positive perceptions about their jobs (such as feeling secure in their positions, perceiving opportunities for advancement, and receiving appropriate guidance in how to complete specific tasks). However, these positive items were not as strongly endorsed as were the more negative aspects of job stress.

The Effects of Job Stresses. The survey results indicated that Job Stresses affect both Physical Health and General Well-being. Individuals who experienced greater job pressures reported more ailments and more bad headaches, took more sickdays, and had worse overall scores on a General Health Questionnaire. They also considered themselves more out of shape and less physically well than the rest of the sample. By contrast, those who felt the most secure about their jobs had the fewest bad headaches and felt least out of shape. Perceived Job Stresses also had impacts on individuals' mental health. Those who experienced higher job pressures were less satisfied with their lives in general, were less relaxed and less able to enjoy themselves, and felt more harried than the rest of the sample. Not surprisingly, they also reflected lower levels of job satisfaction than did the rest of the sample.

The Value of Workday Restoratives. Participants' responses indicated that a variety of brief, non-work-related involvements were at least moderately effective in relieving job stress. Some of these restorative experiences involved familiar, "nonproductive" activities. For example, "talking to others" was considered generally restorative. However, this activity was less effective for employees who perceived their jobs as being more stressful and was also associated with lower levels of General Well-being. "Taking some personal time during work" was also considered moderately restorative. People who were more frustrated with their jobs found focusing on their own interests during work even more helpful than others did. Getting some physical exercise during the workday was also considered moderately restorative by all the participants, and even more so by those whose work involved group meetings. It was also associated with fewer health problems and higher levels of General Well-being.

All of the other workday involvements that were considered moderately restorative involved some kind of contact with the outdoors: Walking out on a balcony, eating lunch outdoors, and sitting or eating near a body of water are some examples. A set of environmental contact items (enjoying the view outside, taking a walk outside, and listening to music) also received moderately restorative ratings.

The Importance of Nature Views. The value of nature at the workplace was also examined by exploring the relationship between individuals' contact with the outdoors and their perceptions of job stresses as well as of their own physical and mental well-being. Individuals were asked both where they spent most of their time during work (at their desk, at group meetings, working outdoors, and so on) and what things they could see outside from their desk, if they had one.

The results of this aspect of the study were particularly striking. Employees whose outdoor views included only built components (such as roads or buildings) experienced higher levels of job stress than others did. By contrast, desk workers who could see at least some natural elements outdoors (such as trees and grass) reflected higher job satisfaction levels than did either those with views of built elements outdoors or those with no outdoor views from their desks at all. And both workers with nature views and those who usually worked outdoors reported substantially lower levels of job stress. Furthermore, they reported fewer health problems and higher levels of overall life satisfaction than did the rest of the sample.

Noteworthy Points. This study documents the impacts of job stress on both physical and mental well-being. The results relating to nature views and opportunities to be outside during the workday suggest the multifaceted benefits that can be gained by incorporating easily accessible nature in the work setting.

THE FUNCTIONS OF URBAN NATURE: USES AND VALUES OF DIFFERENT TYPES OF URBAN NATURE SETTINGS

Janet T. Talbot, Lisa V. Bardwell, and Rachel Kaplan (1987)
Journal of Architectural and Planning Research, 4, 47–63

Overview. As a follow-up to the Bardwell (1985) study, residents of a similar townhouse complex, which also had many and varied outdoor settings within and adjacent to it. were interviewed about their perceptions and uses of nearby nature. In addition to the original set of questions, the participants were asked to rate the importance of each area and to explain why they valued it. The responses supported and expanded on Bardwell's findings and reflected residents' needs for access to three specific types of nearby-nature areas, which were used differently and served different psychological functions.

Survey/Sample. Photographs depicting nine nearby outdoor areas were used in each interview. Residents were asked the questions used in the Bardwell study; they also rated the importance of each area and described any particular value they attached to any area they rated as being either important or very important.

Eighty-nine residents were interviewed (representing over half of those who were contacted by phone), either in their homes or, if convenient, in a downtown location. The respondents were demographically diverse: Slightly over half were younger than 40 years old, and a similar percentage had children living at home. Many were long-term residents of the complex.

Results

Categories of Nearby Nature. The nine areas studied fell into three distinct categories: Yards (residents' own front and back yards and others' yards), Recreation Spaces (common areas in the complex and the adjacent school playing fields), and Nature Settings (nearby undeveloped scrubby areas and the nearby woodlot and pond areas). Patterns of responses to the ratings and to the open-ended questions were distinct for each of these three groups of areas.

Preferences and Uses. Preferences varied from moderate to extremely high. The Yards were preferred least (with a mean preference of 3.3 on a 5-point scale for the front yards), Recreation Spaces were more highly rated, and the Nature Settings received the highest ratings from the sample (4.6 for the forest, the most preferred setting).

Use frequencies and the types of uses reported were widely varied. Yards were used most frequently, Recreation Spaces were used somewhat less often, and the Nature Settings received the least amount of use. The types of uses reported were predominantly "intentional"; the most frequent uses across all the areas were going for walks, playing sports, gardening, socializing, and enjoying nature-related pursuits. The Yards reflected the greatest variety of uses, including gardening, sitting and relaxing, socializing, and picnics. Recreation Spaces were used primarily for sports, as well as for watching children and taking walks. Nature Settings were used for walking and for nature-related pursuits such as bird watching and picking wildflowers.

Sizes and Size Adequacy. Yards were perceived as relatively small and as being smaller than people would like. The other outdoor areas were considered average to large in size. The Recreation Spaces were considered about right in size, but the Nature Settings seemed a little too small.

Perceived Values. In contrast with the rating results discussed above, the importance ratings did not distinguish between the three perceived categories of nearby nature. At least one specific area in each category received a high (4.0 or better) mean importance rating from the sample. The highest ratings went to both the front and back yards

(4.3 and 4.4, respectively). The explanations of these perceived values most often dealt with the natural beauty that these areas provided, but they served additional functions as well. The Yards met territorial needs (and others' yards prompted feelings of community pride); the Recreation Spaces provided opportunities for outdoor sports; and the Nature Settings were appreciated for the nearness of undeveloped natural settings (which imparted a "rural" character to the area) and for just knowing that they were there.

Interrelationships among Variables. Preference was positively correlated with perceived sizes only for individuals' own front and back yards and for the commons areas in the complex. Regarding size adequacy, the relationship to preference was positive for the very small front yards (the more people liked them, the less inadequate they seemed). For the forest, the pond, and an adjacent scrubby area, the relationship between size adequacy and preference was negative (the more people liked them, the less they seemed "too big" in size).

Noteworthy Points. Again, this study shows that for public open spaces, physical sizes are not related to preference. However, for one's own territory (which in this case includes common spaces within the complex as well as individual yards), both size and size adequacy are positively related to preference.

Three distinct categories of nearby nature were identified: Yards, Recreation Spaces, and Nature Settings. Residents used these areas in many different ways, with little overlap in the types of uses that were typically pursued in each of the three categories of nature areas.

Importance ratings were similar, but the reasons for an area's importance were different for the three categories of areas. Residents perceived a special value in living next to a wooded area, as it changed the character of an otherwise urban region to one that seemed more rural.

Ed. note. Portions of this study are discussed in Talbot, J. F., & Bardwell, L. V. (1989), *Making "open spaces" that work: Research and guidelines for natural areas in medium-density housing.* In G. Hardie (Ed.), *Changing paradigms.*

APPENDIX D NEEDS AND FEARS: THE RESPONSE TO TREES AND NATURE IN THE INNER CITY.

Janet F. Talbot and Rachel Kaplan (1984)
Journal of Arboriculture, 10, 222–228

Overview. Inner-city residents rated photographs of urban nature settings for preference and indicated what specific physical features they liked and disliked. Residents expressed strong concerns for safety and perceived overgrown settings as providing potential hiding spaces for attackers. Despite these fears, they indicated that nature contacts were highly valued in their daily lives and that being in natural surroundings prompted a variety of strongly positive responses.

Survey/Sample. The survey included open-ended questions asking people how important nearby nature was in their daily lives and what kinds of feelings they had while in natural surroundings. Some demographic information was also obtained. The sample consisted of 97 residents of inner-city Detroit neighborhoods, who were interviewed in their homes. The sample was black, low-to-moderate-income, and lived in stable neighborhoods surrounded by commercial and industrial areas.

Results

Personal Responses to Nature Contact. Comments indicated very positive feelings about being in nearby-nature areas: 85% found this Relaxing (restful, soothing); 76% found it Enjoyable (pleasant, good feelings); 40% found it a chance to Escape Worries (get rid of worries, escape city pressures, forget tensions); and 32% indicated it gave them time to Think (let thoughts wander).

Importance of Nature Contacts. Participants were asked both about the frequency and about the importance of their own contacts with nature. The responses to these two questions were combined into one variable, Perceived Value. Only 18% of the participants gave nature involvement a Low value (neither a frequent concern nor extremely important); 4% indicated a Moderate value (not extremely important to them, but a daily interest); for 30% nearby nature held a High value (very important and a frequent involvement); for the remaining 4% it was an integral part of their Daily Life (very important and a daily involvement).

Predictors of Personal Response and Importance Ratings. Demographic variables made few differences in these ratings. People from households with children more often mentioned that being outdoors gave them time to Think. And elderly respondents less often described their feelings while outdoors as Enjoyable. None of the six demographic variables (age, sex, income, urban/rural background, length of residence in Detroit, or household type) made a significant difference in the Perceived Value of nature contacts.

Noteworthy Points. These results clearly show high perceived value of nearby nature among a sample of urban blacks, whose opportunities to enjoy nature are limited and whose needs for nearby nature have been questioned by other researchers. There were no significant differences by income or background in the importance of nature contacts.

Ed. note. Portions of this study are also discussed in:

Talbot, J. F. (1988). Planning concerns related to urban nature environments: The role of size and other physical features. In J. L. Nasar (Ed.), *Environmental aesthetics: Theory, research, and applications.* New York: Cambridge University Press.

Kaplan, R., & Talbot, J. F. (1988). Ethnicity and preference for natural settings: A review and recent findings. *Landscape and Urban Planning, 15,* 107–117.

Discussion here focuses on the importance of the natural environment. This study is also included in Appendix B, where the focus is on the preference aspects, based on ratings of photographs.

JUDGING THE SIZES OF URBAN OPEN AREAS: IS BIGGER ALWAYS BETTER?

Janet F. Talbot and Rachel Kaplan (1986)
Landscape Journal, 5, 83–92

Overview. This study explored the perceived sizes of different urban nature areas. The results showed that participants judged the relative sizes of areas fairly accurately but that neither perceived nor actual sizes were correlated with the participants' preference ratings for the same areas. Verbal descriptions by the participants of the visual cues that they used in judging sizes and preferences were very similar, however.

Survey/Sample. Participants examined photographs depicting outdoor areas, which varied both in size and in physical appearance. Four black-and-white photographs, each taken from a different vantage point, were used to represent each of the 15 areas examined. The participants sorted the areas into five piles, first according to size, and then according to how well they liked each setting. After each of the two sorting tasks, the participants indicated which specific physical features were useful in evaluating the areas.

The sample consisted of 56 participants, who were approached in a variety of public settings. Males and females were equally represented, as were both homeowners and renters; ages were quite varied.

Results

Perceived Sizes and Size Cues. The correlation between real and rated sizes across the 15 areas was .59 (significant at $p < .05$). Some of the cues that people mentioned as helping them judge sizes were consistent with the average size ratings given to the areas: Settings with at least some open grassy space looked larger to people, whereas areas that included more built components and areas edged by buildings or fences were rated as smaller. However, other physical features served as inconsistent size cues. Features such as paths, mowed areas, and large trees or large numbers of trees were mentioned as affecting size judgments but were not visibly different in areas that were rated as large or small than they were in the rest of the areas studied.

Relationship between Size and Preference. The correlation between size and preference ratings across the 15 areas was not significant (nor was the correlation between actual sizes and preference ratings). Despite the independence of the two ratings, the participants' descriptions of features that helped them judge sizes and features that helped them judge preferences were very similar. Seven of the eight specific cues that were mentioned as making places look large were also mentioned as making places more highly preferred.

Noteworthy Points. This is one of the very few studies to deal with size issues explicitly. For the range of sizes included, the results showed that preferences are independent of physical sizes.

The similarity in the verbal explanations that people gave for their size and preference ratings suggests that people think that large spaces and preferred settings share similar characteristics, even though the independence of the two rating sets indicates that this is not the case. The article includes some specific design guidelines for making either small or large places feel spacious.

REFERENCES

Asterisk designates study that appears in Appendix B or D.

Alexander, C., Ishikawa, S., & Silverstein, M. (1977). *A pattern language.* New York: Oxford University Press.

*Anderson, E. (1978). *Visual resource assessment: Local perceptions of familiar natural environments.* Unpublished doctoral dissertation, University of Michigan, Ann Arbor.

Appleton, J. (1975). *The experience of landscape.* London: Wiley.

Balling, J. D., & Falk, J. H. (1982). Development of visual preference for natural environments. *Environment and Behavior, 14,* 5-28.

*Bardwell, L. V. (1985). *Nature around the corner: Preference and use of nearby natural areas in the urban setting.* Unpublished master's thesis, University of Michigan, Ann Arbor.

Berris, C. R. (1988). *Suburban housing environments: Residential preferences and satisfactions.* Manuscript submitted for publication.

Brickman, P. (1978). Is it real? *Journal of Experiential Learning and Simulation, 2,* 39–53.

Brickman, P., & Campbell, D. T. (1971). Hedonic relativism and planning: The good society. In M. H. Appley (Ed.), *Adaptation level theory.* New York: Academic Press.

Bruner, J. S., & Postman, L. (1949). On the perception of incongruity: A paradigm. *Journal of Personality, 18,* 206–223.

Buhyoff, G. J., Leuschner, W. A., & Wellman, J. D. (1979). Aesthetic impacts of southern pine beetle damage. *Journal of Environmental Management, 8,* 261–267.

Buhyoff, G. J., Wellman, J. D., Harvey, H., & Fraser, R. A. (1978). Landscape architects' interpretations of people's landscape preferences. *Journal of Environmental Management, 6,* 255–262.

Burton, L. M. (1981). *A critical analysis and review of the research on Outward Bound and related programs.* Unpublished doctoral dissertation, Rutgers University, New Brunswick, NJ.

Cawte, J. E. (1967). Flight into the wilderness as a psychiatric syndrome. *Psychiatry, 30,* 149–161.

Clark, R. N., & Stankey, G. H. (1979). *The recreation opportunity spectrum: A framework for planning, management, and research.* USDA Forest Service General Technical Report PNW-98.

Coeterier. J. F. (1983). A photo validity test. *Journal of Environmental Psychology, 3,* 315–323.

Cohen, S., & Lezak, A. (1977). Noise and inattentiveness to social cues. *Environment and Behavior, 9,* 559–572.

Cohen, S., & Spacapan, S. (1978). The aftereffects of stress: An attentional interpretation. *Environmental Psychology and Nonverbal Behavior, 3,* 43–57.

Cooper-Marcus, C., & Sarkissian, W. (1986). *Housing as if people mattered.* Berkeley: University of California Press.

Cordell, H. K., & Hendee, J. C. (1982). *Renewable resources recreation in the United States: Supply, demand, and critical policy analysis.* Washington, DC: American Forestry Association.

Csikszentmihalyi, M. (1975). *Beyond boredom and anxiety.* San Francisco: Jossey-Bass.

Csikszentmihalyi, M. (1978). Intrinsic rewards and emergent motivation. In M. R. Lepper & D. Greene (Eds.), *The hidden costs of reward: New perspectives on the psychology of human motivation.* Hillsdale, NJ: Erlbaum.

Cullen, G. (1961). *Townscape.* New York: Reinhold.

Daniel, T. C., & Boster, R. S. (1976). *Measuring landscape aesthetics: The scenic beauty estimation method.* USDA Forest Service Research Paper RM-167.

Day, H. (1967). Evaluation of subjective complexity, pleasingness, and interestingness for a series of random polygons varying in complexity. *Perception and Psychophysics, 2,* 281–286.

Donnerstein, E., & Wilson, D. W. (1976). Effects of noise and perceived control on ongoing and subsequent aggressive behavior. *Journal of Personality and Social Psychology, 34,* 774–781.

Driver, B. L. (1972). Potential contributions of psychology to recreation resource management. In J. F. Wohlwill & D. H. Carson (Eds.), *Environment and the social scien-*

ces. Washington, DC: American Psychological Association.

Driver, B. L., & Brown, P. J. (1978). The opportunity spectrum concept and behavioral information in outdoor recreation resource supply inventories: A rationale. In *Integrated inventories of renewable natural resources: Proceedings of the workshop.* USDA Forest Service General Technical Report RM-55.

Driver, B. L., & Knopf, R. C. (1976). Temporary escape: One product of sport fisheries management. *Fisheries,* 1(2), 24–29.

Driver, B. L., Nash, R., & Haas, G. (1987). Wilderness benefits: State of knowledge. In R. C. Lucas (Ed.), *Proceedings of the National Wilderness Research Conference.* Ft. Collins, CO: USDA Forest Service Intermountain Experiment Station.

Dwyer, J. F., Hutchinson, R., & Wendling, R. C. (1981, October 26). *Participation in outdoor recreation by black and white Chicago households.* Paper presented at the National Recreation and Park Association Symposium on Leisure Research, Minneapolis.

*Ellsworth, J. C. (1982). *Visual assessment of rivers and marshes: An examination of the relationship of visual units, perceptual variables, and preference.* Unpublished master's thesis, Utah State University, Logan.

Flinn, N. (1985). *The prison garden book.* Burlington, VT: National Gardening Association.

Fly, J. M. (1986). *Nature, outdoor recreation, and tourism: The basis for regional population growth in northern lower Michigan.* Unpublished doctoral dissertation, University of Michigan, Ann Arbor.

Fox, T., Koeppel, I., & Kellam, S. (1985). *Struggle for space: The greening of New York City, 1970–1984.* Washington, DC: Island Press.

Francis, M. (1987a). The garden inside. *UC Davis Magazine,* Fall, 20–23.

Francis, M. (1987b). Urban open spaces. In E. H. Zube & G. T. Moore (Eds.), *Advances in environment, behavior, and design* (Vol. 1). New York: Plenum.

Francis, M., Cashdan, L., & Paxson, L. (1984). *Community open spaces.* Washington, DC: Island Press.

Francis, M., & Hester, R. T., Jr. (1987). *Meanings of the garden: Conference Proceedings.* Davis: Center for Design Research, University of California.

*Frey, J. E. (1981). *Preferences, satisfactions, and the physical environments of urban neighborhoods.* Unpublished doctoral dissertation, University of Michigan, Ann Arbor.

Fried, M. (1982), Residential attachment: Sources of residential and community satisfaction. *Journal of Social Issues, 38,* 107–120.

Fried, M. (1984). The structure and significance of community satisfaction. *Population and Behavior, 7,* 61–86.

Frisbie, R. (1969). *It's a wise woodsman who knows what's biting him.* New York: Doubleday.

*Gallagher, T. J. (1977) *Visual preference for alternative natural landscapes.* Unpublished doctoral dissertation, University of Michigan, Ann Arbor.

Getz, D. A., Karow, A., & Kielbaso, J. J. (1982). Inner city preference for trees and urban forestry programs. *Journal of Arboriculture, 8,* 258–263.

Gibson, J. J. (1979). *The ecological approach to visual perception.* Boston: Houghton Mifflin.

*Gimblett, R. H., Itami, R. M., & Fitzgibbon, J. E. (1985). Mystery in an information processing model of landscape preference. *Landscape Journal, 4,* 87–95.

Gregory, R. L. (1966). *Eye and brain: The psychology of seeing.* New York: McGraw Hill.

*Hammitt, W. E. (1978). *Visual and user preference for a bog environment.* Unpublished doctoral dissertation, University of Michigan, Ann Arbor.

*Hammitt, W. E. (1979). Measuring familiarity for natural environments through visual images. In *Proceedings of Our National Landscape Conference* (pp. 217–226). USDA Forest Service General Technical Report PSW-35.

*Hammitt, W. E. (1981). The familiarity-preference component of on-site recreational experiences. *Leisure Sciences, 4,* 177–193.

Hammitt, W. E. (1982). Cognitive dimensions of wilderness solitude. *Environment and Behavior, 14,* 478–493.

Hammitt, W, E. (1987). Visual recognition capacity during outdoor recreation experiences. *Environment and Behavior, 19,* 651–672.

Hammitt, E. W., & Brown, G. F., Jr. (1984). Functions of privacy in wilderness environments. *Leisure Sciences, 6,* 151–166.

Hanson, R. A. (1973). Outdoor Challenge and mental

health. *Naturalist, 24,* 26–30.

Hartig, T., Mang, M., & Evans, G. W. (1987, September). Perspectives on wilderness: Testing the theory of restorative environments. In T. Easley & J. Passineau (Co-Chairs), *The use of wilderness for personal growth, therapy, and education.* Symposium conducted at the Fourth World Wilderness Congress, Estes Park, CO.

Hayward, J. (1989). Urban parks: Research, planning and social change. In I. Altman & E. Zube (Eds.), *Public space and places.* New York: Plenum.

Hebb, D. O. (1951). The role of neurological ideas in psychology. *Journal of Personality, 20,* 391–55.

*Herbert, E. J. (1981). *Visual resource analysis: Prediction and preference in Oakland County, Michigan.* Unpublished master's thesis, University of Michigan, Ann Arbor.

*Herzog, T. R. (1984). A cognitive analysis of preference for field-and-forest environments. *Landscape Research, 9,* 10–16.

*Herzog, T. R. (1985). A cognitive analysis of preference for waterscapes. *Journal of Environmental Psychology, 5,* 225–241.

*Herzog, T. R. (1987). A cognitive analysis of preference for natural environments: Mountains, canyons, and deserts. *Landscape Journal, 6,* 140–152.

*Herzog, T. R. (1989). *A cognitive analysis of preference for urban nature. Journal of Environmental Psychology, 9.*

*Herzog, T. R., Kaplan, S. & Kaplan, R. (1976). The prediction of preference for familiar urban places. *Environment and Behavior, 8,* 627–645.

*Herzog, T. R., Kaplan, S., & Kaplan, R. (1982). The prediction of preference for unfamiliar urban places. *Population and Environment, 5,* 43–59.

Herzog, T. R., & Smith, G. A. (1988). Danger, mystery, and environmental preference. *Environment and Behavior, 20,* 320–344.

Hidi, S., & Bard, W. (1986). Interestingness: A neglected variable in discourse processing. *Cognitive Science, 10,* 179–194.

Hodgson, R. W., & Thayer, R. L., Jr. (1980). Implied human influence reduces landscape beauty. *Landscape Planning, 7,* 171–179.

Hollender, J. W. (1977). Motivational dimensions of leisure experience. *Journal of Leisure Research, 9,* 133–141.

Hubbard, H. V., & Kimball, T. (1917). *An introduction to the study of landscape design*. New York: Macmillan.

*Hudspeth, T. R. (1982). *Visual preference as a tool for citizen participation: A case study of urban waterfront revitalization in Burlington, Vermont*. Unpublished doctoral dissertation, University of Michigan, Ann Arbor.

*Hudspeth, T. R. (1986). Visual preference as a tool for facilitating citizen participation in urban waterfront revitalization. *Journal of Environmental Management, 23*, 373–385.

Ittelson, W. H., Proshansky, H. M., Rivlin, L. G., & Winkel, G. H. (1974). *An introduction to environmental psychology*. New York: Holt, Rinehart & Winston.

James, W. (1892). *Psychology: The briefer course*. New York: Holt.

James, W. (1902). *The varieties of religious experience*. New York: Holt.

Kaplan, R. (1972). The dimensions of the visual environment: Methodological considerations. In W. J. Mitchell (Ed.), *Environmental design: Research and practice*. Washington, DC: EDRA.

Kaplan, R. (1973a). Some psychological benefits of gardening. *Environment and Behavior, 5*, 145–152.

*Kaplan, R. (1973b). Predictors of environmental preference: Designers and "clients." In W. F. E. Preiser (Ed.), *Environmental design research*. Stroudsburg, PA: Dowden, Hutchinson, & Ross.

Kaplan, R. (1974). Some psychological benefits of an Outdoor Challenge Program. *Environment and Behavior, 6*, 101 116.

*Kaplan, R. (1975). Some methods and strategies in the prediction of preference. In E. H. Zube, R. O. Brush, & J. G. Fabos (Eds.), *Landscape assessment: Values, perceptions, and resources*. Stroudsburg, PA: Dowden, Hutchinson, & Ross.

Kaplan, R. (1976). Way-finding in the natural environment. In G. T. Moore & R. G. Golledge (Eds.), *Environmental knowing: Theories, perspectives, and methods*. Stroudsburg, PA: Dowden, Hutchinson, & Ross.

*Kaplan, R. (1977a). Preference and everyday nature: Method and application, In D. Stokols (Ed.), *Perspectives on environment and behavior: Theory, research, and applications*. New York: Plenum.

Kaplan, R. (1977b). Patterns of environmental preference.

Environment and Behavior, 9, 195–216.

Kaplan, R. (1980). Citizen participation in the design and evaluation of a park. *Environment and Behavior, 12,* 494–507.

*Kaplan, R. (1982). Managing greenspace in multiple-family neighborhoods. In *Proceedings of the 1982 Convention of the Society of American Foresters.* SAF Publication 83-04.

*Kaplan, R. (1983). The role of nature in the urban context. In I. Altman & J. F. Wohlwill (Eds.), *Behavior and the natural environment.* New York: Plenum.

Kaplan, R. (1984a). Impact of urban nature: A theoretical analysis. *Urban Ecology, 8,* 189–197.

*Kaplan, R. (1984b). Wilderness perception and psychological benefits: An analysis of a continuing program. *Leisure Sciences, 6,* 271–290.

*Kaplan, R. (1984c). Dominant and variant values in environmental preference. In A. S. Devlin & S. L. Taylor (Eds.), *Environmental preference and landscape management.* New London: Connecticut College.

Kaplan, R. (1984d). Assessing human concerns for environmental decision-making. In S. L. Hart, G. A. Enk, & W. F. Hornick (Eds.), *Improving impact assessment: Increasing the relevance and utilization of technical and scientific information.* Boulder, CO: Westview.

*Kaplan, R. (1985a). Nature at the doorstep: Residential satisfaction and the nearby environment. *Journal of Architectural and Planning Research, 2,* 115–127.

Kaplan, R. (1985b). The analysis of perception via preference: A strategy for studying how the environment is experienced. *Landscape Planning, 12,* 161–176.

*Kaplan, R., & Herbert, E. J. (1987). Cultural and subcultural comparisons in preference for natural settings. *Landscape and Urban Planning, 14,* 281–293.

*Kaplan, R., & Herbert, E. J. (1988). Familiarity and preference: A cross-cultural analysis. In J. L. Nasar (Ed.), *Environmental aesthetics: Theory, research, and applications.* New York: Cambridge University Press.

*Kaplan, R., & Kaplan, S. (1987). The garden as restorative experience: A research odyssey. In M. Francis & R. T. Hester (Eds.), *The meanings of the garden: Conference proceedings.* Davis: Center for Design Research, University of California.

*Kaplan, R., Kaplan, S., & Brown, T. J. (in press). Environmental preference: A comparison of four domains of predictors. *Environment and Behavior.*

*Kaplan, R., & Talbot, J. F. (1988). Ethnicity and preference for natural settings: A review and recent findings. *Landscape and Urban Planning, 15,* 107–117.

Kaplan, S. (1970). The role of location processing in the perception of the environment. In J. Archea & C. Eastman (Eds.), *Edra two.* Stroudsburg, PA: Dowden, Hutchinson, & Ross.

Kaplan, S. (1973). Cognitive maps in perception and thought. In R. M. Downs & D. Stea (Eds.), *Image and environment.* Chicago: Aldine.

Kaplan, S. (1975). An informal model for the prediction of preference. In E. H. Zube, R. O. Brush, & J. G. Fabos (Eds.), *Landscape assessment: Values, perceptions, and resources.* Stroudsburg, PA: Dowden, Hutchinson, & Ross.

Kaplan, S. (1977). Tranquility and challenge in the natural environment. In *Children, nature, and the urban environment.* USDA Forest Service General Technical Report NE-30.

Kaplan, S. (1978). Attention and fascination: The search for cognitive clarity. In S. Kaplan & R. Kaplan (Eds.), *Humanscape: Environments for people.* (Republished, Ann Arbor, MI: Ulrich's, 1982.)

Kaplan, S. (1979a). Concerning the power of content-identifying methodologies. In *Assessing amenity resource values.* USDA Forest Service General Technical Report RM-68.

Kaplan, S. (1979b). Perception and landscape: Conceptions and misconceptions. In *Proceedings of Our National Landscape Conference.* USDA Forest Service General Technical Report PSW-35.

Kaplan, S. (1983). A model of person–environment compatibility. *Environment and Behavior, 15,* 311–332.

Kaplan, S. (1987a). Aesthetics, affect, and cognition: Environmental preferences from an evolutionary perspective. *Environment and Behavior, 19,* 3–32.

Kaplan, S. (1987b). Mental fatigue and the designed environment. In J. Harvey & D. Henning (Eds.), *Public environments.* Washington, DC: EDRA.

Kaplan, S., & Kaplan, R. (1978). *Humanscape: Environ-*

ments for people. (Republished, Ann Arbor, MI: Ulrich's, 1982.)

Kaplan, S., & Kaplan, R. (1982). *Cognition and environment: Functioning in an uncertain world.* New York: Praeger.

*Kaplan, S., & Kaplan, R. (1989). The visual environment: Public participation in design and planning. *Journal of Social Issues, 45,* 59–86.

*Kaplan, S., Kaplan, R., & Wendt, J. S. (1972). Rated preference and complexity for natural and urban visual material. *Perception and Psychophysics, 12,* 354–356.

Kaplan, S., & Talbot, J. F. (1983). Psychological benefits of a wilderness experience. In I. Altman & J. F. Wohlwill (Eds.), *Behavior and the natural environment.* New York: Plenum.

*Kaplan, S., Talbot, J. F., & Kaplan, R. (1988). *Coping with daily hassles: The impact of nearby nature on the work environment.* Project Report. USDA Forest Service, North Central Forest Experiment Station, Urban Forestry Unit Cooperative Agreement 23-85-08.

Kaplan, S., & Wendt, J. S. (1972). Preference and the visual environment: Complexity and some alternatives. In W. J. Mitchell (Ed.), *Environmental design: Research and practice.* Washington, DC: EDRA.

Kellomaki, S., & Savolainen, R. (1984). The scenic value of the forest landscape as assessed in the field and the laboratory. *Landscape Planning, 11,* 97–107.

Keyes, B. E. (1984). *Visual preferences of a forest trail environment.* Unpublished master's thesis, University of Tennessee, Knoxville.

Klausner, S. L. (1971). *On man in his environment.* San Francisco: Jossey-Bass.

Kluckhohn, F. R. (1953). Dominant and variant value orientations. In C. Kluckhohn & H. A. Murray (Eds.), *Personality in nature, society, and culture.* New York: Knopf.

Knopf, R. C. (1983). Recreational needs and behavior in natural settings. In I. Altman and J. F. Wohlwill (Eds.), *Behavior and the natural environment.* New York: Plenum.

Knopf, R. C. (1987). Human behavior, cognition, and affect in the natural environment. In D. Stokols & I. Altman (Eds.), *Handbook of environmental psychology.* New York: Wiley.

Kulik, J. A., Revelle, W. R., & Kulik, C-L. (1970). *Scale construction by hierarchical cluster analysis.* Unpublished manuscript, University of Michigan, Ann Arbor.

Lakoff, G. (1987). *Women, fire, and dangerous things.* Chicago: University of Chicago Press.

*Levin, J. (1977). *Riverside preference: On-site and photographic reactions.* Unpublished master's thesis, University of Michigan, Ann Arbor.

Lewis, C. A. (1972). Public housing gardens: Landscapes for the soul. In *Landscape for living.* Washington, DC: USDA Yearbook of Agriculture.

Lewis, C. A. (1979). Healing in the urban environment: A person/plant viewpoint. *Journal of the American Planning Association, 45,* 330–338.

Lezak, M. D. (1982). The problem of assessing executive functions. *International Journal of Psychology, 17,* 281–297.

Lingoes, J. C. (1972). A general survey of the Guttman-Lingoes nonmetric program series. In R. N. Shepard, A. K. Romney, & S. B. Nerlove (Eds.), *Multidimensional scaling* (Vol. 1). New York: Seminar.

Little, C. E. (1974). The double standard of open space. In J. N. Smith (Ed.), *Environmental quality and social justice in urban America.* Washington, DC: Conservation Foundation.

Lucas, R. C. (1980). *Use patterns and visitor characteristics, attitudes, and preferences in nine wilderness and other roadless areas.* USDA Forest Service Research Paper INT-253.

Lynch, K. (1960). *The image of the city.* Cambridge, MA: MIT Press.

Mang, M. (1984). *The restorative effects of wilderness backpacking.* Unpublished doctoral dissertation, University of California, Irvine.

Maslow, A. (1962). *Toward a psychology of being.* Princeton, NJ: Van Nostrand.

*Medina, A. Q. (1983). *A visual assessment of children's and environmental educators' urban residential preference patterns.* Unpublished doctoral dissertation, University of Michigan, Ann Arbor.

Midgley, M. (1978). *Beast and Man: The roots of human nature.* Ithaca, NY: Cornell University Press.

*Miller, P. A. (1984). *Visual preference and implications for coastal management: A perceptual study of the*

British Columbia shoreline. Unpublished doctoral dissertation, University of Michigan, Ann Arbor.

Milner, P. M. (1957). The cell assembly: Mark II. *Psychological Review. 64,* 242–252.

Moore, E. O. (1981). A prison environment's effect on health care service demands. *Journal of Environmental Systems, 11,* 17–34.

Mueller, E., Kennedy, J. M., & Tanimoto, S. M. (1972). Inherent perceptual motivation and the discovery of structure. *Journal of Structural Learning, 3,* 1–6.

Murphy, G. L., & Medin, D. L. (1985). The role of theories in conceptual coherence. *Psychological Review, 92,* 289–316.

Naimark, S. (1982). *A handbook of community gardening.* New York: Scribner.

National Gardening Association (1985). *Special report on community gardening in the U.S.* Burlington, VT: National Gardening Association.

National Urban Forest Forum (1988, January/February). *Shedding a few tears.* Washington, DC: American Forestry Association.

Newman, O. (1972). *Defensible space.* New York: Macmillan.

Newman, R. S. (1980). Alleviating learned helplessness in a wilderness setting: An application of attribution theory to Outward Bound. In L. J. Fyans, Jr. (Ed.), *Achievement motivation: Recent trends in theory and research.* New York: Plenum.

Peterson, G. L. (1977). Recreational preferences for urban teenagers: The influence of cultural and environmental attributes. In *Children, nature, and the urban environment.* USDA Forest Service General Technical Report NE-30.

Pitt, D. G., & Zube, E. H. (1979). The Q-sort method: Use in landscape assessment research and landscape planning. In *Proceedings of Our National Landscape Conference.* USDA Forest Service General Technical Report PSW-35.

Pitt, D. G., & Zube, E. H. (1987). Management of natural environments. In D. Stokols & I. Altman (Eds.), *Handbook of environmental psychology.* New York: Wiley.

Porter, M. V. (1987). *The role of spatial quality and familiarity in determining countryside landscape preference.* Unpublished master's thesis. University of Washington, Seattle.

Posner, M. I. (1986). Empirical studies of prototypes. In C. Craig (Ed.), *Noun classes and categorization.* Amsterdam: John Benjamins.

Postman, N. (1985). *Amusing ourselves to death.* New York: Penguin Books.

Robinson, G. O. (1975). *The Forest Service.* Baltimore, MD: Resources for the Future.

Rosch, E. (1978). Principles of categorization. In E. Rosch & B. B. Lloyd (Eds.), *Cognition and categorization.* Hillsdale, NJ: Erlbaum.

Rossman, B. B., & Ulehla, Z. J. (1977). Psychological reward values associated with wilderness use. *Environment and Behavior, 9,* 41–66.

Rowntree, R. A. (1988). Ecology of the urban forest: Introduction to Part III. *Landscape and Urban Planning, 15,* 1–10.

Schauman, S. (1986). Countryside landscape visual assessment. In R. C. Smardon, J. F. Palmer, & J. P. Felleman (Eds.), *Foundations for visual project analysis.* New York: Wiley.

Schauman, S. (1988a). Scenic value of countryside landscapes to local residents: A Whatcom County, Washington, case study. *Landscape Journal, 7,* 40–46.

Schauman, S. (1988b). Countryside scenic assessment: Tools and an application. *Landscape and Urban Planning, 15,* 227–239.

Schauman, S., & Pfender, M. (1982). *An assessment procedure for countryside landscapes.* Seattle: Department of Landscape Architecture, University of Washington.

Schroeder, H. W. (1983). Variations in the perception of urban forest recreation sites. *Leisure Sciences, 5,* 221–230.

Schroeder, H. W. (1984). Environmental perception rating scales: A case for simple methods of analysis. *Environment and Behavior, 16,* 573–598.

Schroeder, H. W. (1988). Environment, behavior, and design research on urban forests. In E. H. Zube & G. T. Moore (Eds.), *Advances in environment, behavior, and design* (Vol. 2). New York: Plenum.

Schroeder, H. W., & Green, T. L. (1985). Public preference for tree density in municipal parks. *Journal of Arboriculture, 11,* 272–277.

Seymour, J. (1979). *The self-sufficient gardener.* New York: Doubleday.

Shafer, E. L., & Mietz, J. (1969). Aesthetic and emotional

experiences rate high with northeast wilderness hikers. *Environment and Behavior, 1,* 187–197.

Shafer, E. L., & Tooby, M. (1973). Landscape preferences: An international replication. *Journal of Leisure Research, 5,* 60–65.

Sherrod, D. R., & Downs, R. (1974). Environmental determinants of altruism: The effects of stimulus overload and perceived control on helping. *Journal of Experimental Social Psychology, 10,* 468–479.

Shuttleworth, S. (1980). The use of photographs as an environmental presentation medium in landscape studies. *Journal of Environmental Management, 11,* 61–76.

Simonds, O. C. (1920). *Landscape-gardening.* New York: Macmillan.

Smardon, R. C. (1988). Perception and aesthetics of the urban environment: Review of the role of vegetation. *Landscape and Urban Planning, 15,* 85–106.

Stankey, G. H. (1980). *A comparison of carrying capacity perceptions among visitors to two wildernesses.* USDA Forest Service Research Paper INT-242.

*Stino, L. (1983). *A visual preference study of urban outdoor spaces in Egypt.* Unpublished doctoral dissertation, University of Michigan, Ann Arbor.

Stone, J. (1971). *The monster at the end of this book.* New York: Western.

Stribley, K. A. (1976). Looking back on public housing. Unpublished master's thesis, University of Michigan, Ann Arbor.

Stringer, P. (1975). The natural environment. In D. Canter & P. Stringer (Eds.), *Environmental interaction.* London: Surrey University Press.

*Talbot, J. F. (1982). Zoning reconsidered: The impact of environmental aesthetics in urban neighborhoods. In P. Bart, A. Chen, & G. Francescato (Eds.), *EDRA 13: Knowledge and design* (pp. 154–159). Washington, DC: Environmental Design Research Association.

*Talbot, J. F. (1988). Planning concerns related to urban nature environments: The role of size and other physical features. In J. L. Nasar (Ed.), *Environmental aesthetics: Theory, research, and applications.* New York: Cambridge University Press.

*Talbot, J. F., & Bardwell, L. V. (1989). *Making "open spaces" that work: Research and guidelines for natural areas in medium-density housing.* In G. Hardie (Ed.),

Changing paradigms. Washington, DC: Environmental Design Research Association.

*Talbot, J. F., Bardwell, L. V., & Kaplan, R. (1987) The functions of urban nature: Uses and values of different types of urban nature settings. *Journal of Architectural and Planning Research, 4,* 47–63.

*Talbot, J. F., & Kaplan, R. (1984). Needs and fears: The response to trees and nature in the inner city. *Journal of Arboriculture, 10,* 222–228.

*Talbot, J. F., & Kaplan, R. (1986). Judging the sizes of urban open areas: Is bigger always better? *Landscape Journal, 5,* 83–92.

Talbot, J. F., & Kaplan, S. (1986). Perspectives on wilderness: Re-examining the values of extended wilderness experiences. *Journal of Environmental Psychology, 6,* 177–188.

Thomas, J. C. (1977). *Cognitive psychology from the perspective of wilderness survival.* IBM Research Report RC-6647, No. 28603. Yorktown Heights, NY: IBM.

Tversky, B., & Hemenway, K. (1983). Categories of environmental scenes. *Cognitive Psychology, 15,* 121–149.

Tyll, L. T. (1988). *Contemporary approaches to stress management.* Unpublished manuscript, University of Michigan, Ann Arbor.

*Ulrich, R. S. (1974). *Scenery and the shopping trip: The roadside environment as a factor in route choice.* Unpublished doctoral dissertation, University of Michigan, Ann Arbor.

Ulrich, R. S. (1979). Visual landscapes and psychological well-being. *Landscape Research, 4,* 17–23.

Ulrich, R. S. (1983). Aesthetic and affective response to natural environment. In I. Altman & J. F. Wohlwill (Eds.), *Behavior and the natural environment.* New York: Plenum.

Ulrich, R. S. (1984). View through a window may influence recovery from surgery. *Science, 224,* 420–421.

Ulrich, R. S. (1986). Human responses to vegetation and landscapes. *Landscape and Urban Planning, 13,* 29–44.

Ulrich, R. S., & Addoms, D. L. (1981). Psychological and recreational benefits of a residential park. *Journal of Leisure Research, 13,* 43–65.

U.S. Bureau of Land Management (1980). *Visual resource management program.* Washington, DC: U.S. Department of the Interior.

U.S. Forest Service (1974). The visual management system. In *National Forest landscape management* (Vol. 2). Washington, DC: U.S. Department of Agriculture.

Verderber, S. (1986). Dimensions of person–window transactions in the hospital environment. *Environment and Behavior, 18,* 450–466.

Vitz, P. C. (1966). Affect as a function of stimulus variation. *Journal of Experimental Psychology, 71,* 74–79.

Vining, J., & Stevens, J. J. (1986). The assessment of landscape quality: Major methodological considerations. In R. C. Smardon, J. F. Palmer, & J. P. Felleman (Eds.), *Foundations for visual project analysis.* New York: Wiley.

Washburne, R., & Wall, P. (1980). *Black–white differences in outdoor recreation.* USDA Forest Service Research Paper INT-249.

Watzlawick, P., Weakland, J. H., & Fisch, R. (1974). *Change: Principles of problem formation and problem resolution.* New York: Norton.

*Weber, W. W. (1980). *Comparison of media for public participation in natural environmental planning.* Unpublished doctoral dissertation, University of Michigan, Ann Arbor.

Wecker, S. C. (1964). Habitat selection. *Scientific American, 211(4),* 109–116.

West, M. J. (1986). *Landscape views and stress responses in the prison environment.* Unpublished master's thesis, University of Washington, Seattle.

Widmar, R. (1983). *Preferences for multiple-family housing in small city neighborhoods.* Unpublished doctoral dissertation, University of Michigan, Ann Arbor.

Wohlwill, J. F., & Harris, G. (1980). Response to congruity or contrast for manmade factors in natural-recreation settings. *Leisure Sciences, 3,* 349–365.

*Woodcock, D. M. (1982). *A functionalist approach to environmental preference.* Unpublished doctoral dissertation, University of Michigan, Ann Arbor.

*Yang, B. (1988). *A cross-cultural comparison of preference for Korean, Japanese, and western landscape styles.* Unpublished doctoral dissertation, University of Michigan, Ann Arbor.

Zube, E. H. (1984). Themes in landscape assessment theory. *Landscape Journal, 3,* 104–110.

Zube, E. H., & Mills, L. V., Jr. (1976). Cross-cultural explorations in landscape perception. In E. H. Zube (Ed.), *Studies in landscape perception*. Institute for Man and Environment, University of Massachusetts, Amherst.

Zube, E. H., & Pitt, D. G. (1981). Cross-cultural perceptions of scenic and heritage landscapes. *Landscape Planning, 8*, 69–87.

Zube, E. H., Pitt, D. G., & Anderson, T. W. (1975). Perception and prediction of scenic resource values of the Northeast. In E. H. Zube, R. O. Brush, & J. G. Fabos (Eds.), *Landscape assessment: Values, perceptions, and resources*. Stroudsburg, PA: Dowden, Hutchinson, & Ross.

Zube, E. H., Pitt, D. G., & Evans, G. W. (1983). A lifespan developmental study of landscape assessment. *Journal of Environmental Psychology, 3*, 115–128.

Zube, E. H., Simcox, D. E., & Law, C. S. (1987). Perceptual landscape simulations: History and prospect. *Landscape Journal, 6*, 62–80.

AUTHOR INDEX

SUBJECT INDEX